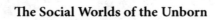

The Social Worlds of the Unborn

DOI: 10.1057/9781137310729

Also by Deborah Lupton

MORAL THREATS AND DANGEROUS DESIRES: AIDS in the News Media

THE FIGHT FOR PUBLIC HEALTH: Principles and Practice of Media Advocacy (with Simon Chapman)

THE IMPERATIVE OF HEALTH: Public Health and the Regulated Body

FOOD, THE BODY AND THE SELF

THE NEW PUBLIC HEALTH: Health and Self in the Age of Risk (with Alan Petersen)

TELEVISION, AIDS AND RISK: A Cultural Studies Approach to Health Communication (with John Tulloch)

CONSTRUCTING FATHERHOOD: Discourses and Experiences (with Lesley Barclay)

THE EMOTIONAL SELF: A Sociocultural Exploration

RISK (2nd edition)

RISK AND SOCIOCULTURAL THEORY: New Directions and Perspectives (edited)

RISK AND EVERYDAY LIFE (with John Tulloch)

MEDICINE AS CULTURE: Illness, Disease and the Body (3rd edition)

FAT

THE UNBORN HUMAN (*edited*)

DOI: 10.1057/9781137310729

palgrave▶pivot

The Social Worlds
of the Unborn

Deborah Lupton
University of Sydney, Australia

palgrave
macmillan

DOI: 10.1057/9781137310729

First published 2013 by
PALGRAVE MACMILLAN

Palgrave Macmillan in the UK is an imprint of Macmillan Publishers Limited,
registered in England, company number 785998, of Houndmills, Basingstoke,
Hampshire RG21 6XS.

Palgrave Macmillan in the US is a division of St Martin's Press LLC,
175 Fifth Avenue, New York, NY 10010.

Palgrave Macmillan is the global academic imprint of the above companies
and has companies and representatives throughout the world.

Palgrave® and Macmillan® are registered trademarks in the United States,
the United Kingdom, Europe and other countries

ISBN: 978-1-137-31073-6 EPUB
ISBN: 978-1-137-31072-9 PDF
ISBN: 978-1-137-31071-2 Hardback

This book is printed on paper suitable for recycling and made from fully
managed and sustained forest sources. Logging, pulping and manufacturing
processes are expected to conform to the environmental regulations of the
country of origin.

A catalogue record for this book is available from the British Library.

A catalog record for this book is available from the Library of Congress.

www.palgrave.com/pivot

DOI: 10.1057/9781137310729

For my daughters Phoebe and Miranda

DOI: 10.1057/9781137310729

Contents

DOI: 10.1057/9781137310729

DOI: 10.1057/9781137310729

palgrave▶pivot

www.palgrave.com/pivot

Introduction

Abstract: *Here I provide an overview of the main themes and concerns of the book, noting that it is about the ways in which we conceptualise, visualise and treat human embryos and foetuses across a range of social and historical contexts. The book will address such topics as pregnancy, prenatal testing and obstetric ultrasound, IVF services, news and social media portrayals of the unborn, abortion politics and human embryonic stem cell research. The theoretical orientation of the book is introduced and the notion that the unborn move across and between social worlds is explained. The concept of the 'assemblage' is employed to describe the ways in which the unborn are configured via discourses, practices, material objects and bodies. The contents of the rest of the chapters in the book are briefly reviewed.*

Key words: embryos; foetuses; the unborn; sociocultural analysis; social theory; assemblages

Lupton, Deborah. *The Social Worlds of the Unborn.* Basingstoke: Palgrave Macmillan, 2013. DOI: 10.1057/9781137310729.

Obstetric ultrasound images on t-shirts, advertisements and customised Christmas cards, vintage anatomical drawings and models of pregnant women and foetuses, foetus dolls, baby shower cakes depicting foetuses, pregnancy handbooks, magazines and websites, preserved dead embryos and foetuses, memorial YouTube videos for lost pregnancies, anti-abortion films, coffee table books and documentaries featuring beautiful intra-uterine photographic images, prenatal testing and screening technologies, *in vitro* fertilisation (IVF) treatments and decisions over what to do with the surplus embryos produced via these treatments, debates in parliaments about regulating abortion and human embryonic stem cell research. What do all of these have in common? These images, bodies, discourses, debates, practices, technologies and objects act to configure and give meaning to unborn entities in diverse and dynamic ways.

Human unborn organisms – embryos and foetuses – have become powerful and suggestive figures in contemporary scientific, legal, philosophical and popular arenas. They are entities invested with many different and often contrasting meanings and emotions, all of which are highly contextual and constantly shifting in response to changes in medical science, technologies and popular representations. Before the advent of scientific medicine, the unborn had to be imagined, brought into being in cultural understandings. The origin narratives constructed around embryos and foetuses in various cultures and religions represent a rich heterogeneity of meaning that relied on myth and metaphor to explain the development of the unborn and their character and relationship with the maternal body (Conklin and Morgan 1996, Duden 1993, Law and Sasson 2009). The increasing medicalisation and politicisation of the unborn across a range of cultural and geographical contexts worldwide, in concert with the growing use of visualising technologies to represent their outward appearance, has led to unborn entities being framed in limited ways that are more literal, related to their apparent health, gender, stage of development and so on. In cultures where biomedical understandings of the unborn have dominated, such as western cultures, the meanings associated with embryos and foetuses have become more fixed and concrete, and thus less ambiguous (Law and Sasson 2009).

For such tiny organisms, the unborn bear an enormous ideological, political, moral, ethical and affective weight. Embryos and foetuses have gained increasing visibility in the public domain to the point that they have become fetishised cultural icons and the subject of fervent contestation over their meanings and ontologies. Part of this process has been

DOI: 10.1057/9781137310729

the infantilising of the unborn, which has reached into the earliest stage of embryonic development so that even new clusters of cells are now frequently referred to as 'babies'. These developments have significant implications for the ways in which legislation is created and enforced in relation to pregnant women and their unborn and the relative rights and responsibilities that are accorded to women in relation to the unborn. Current discourses on pregnancy represent the pregnant woman as custodian of her precious and vulnerable 'baby' and frequently privilege the unborn's needs and rights over those of the woman. The state of pregnancy has become a highly public experience, with women's 'baby bumps' scrutinised for their shape and size and their consumption and comportment of their bodies under surveillance from others and judged on whether they conform to expectations about appropriate pregnant behaviour.

This intensification of focus on unborn entities and their welfare is predominantly due to the creation and introduction of technologies that are able to create, visualise, medically test and screen unborn bodies and treat them for health conditions. Medical care has become oriented towards a 'foetus-centred' approach (Beynon-Jones 2012). Developments in foetal surgery, in which the foetus is operated upon while still *in utero*, position this entity in a particular way: as a 'patient' in its own right. Improvements in care techniques for prematurely born infants at earlier stages of development now mean that the distinction between the 'foetus' and the 'infant' has changed. Infants born as early as 22 weeks of gestation can now be kept alive for a time in specialist neonatal intensive care units (although the vast majority of these infants eventually die or suffer permanent disabilities as a result of extreme prematurity). Other developments in medical technologies directed at the testing or screening of embryos and foetuses, including amniocentesis, chorionic villus sampling, various maternal blood tests and preimplantation genetic diagnosis (PGD), have contributed to the medicalisation and surveillance of the unborn.

Such visualising technologies as the photographic images of embryos and foetuses produced for magazines and coffee table books by photojournalist Lennart Nilsson from the middle of the twentieth century onwards, together with obstetric ultrasound (also referred to as sonography), which began to be used on pregnant women in the 1970s, have contributed to the contemporary imagery of unborn entities: their status as visible and knowable, amenable to monitoring,

DOI: 10.1057/9781137310729

measuring and regulation. More recently, computer-generated images that use a combination of inter-uterine photographs, ultrasound images and preserved anatomical specimens have been able to visually portray the unborn body at each stage of development in fine detail. The latest ultrasound technologies are now able to visualise and monitor the unborn from very early stages of development and in three-dimensions (3D) and four-dimensions (4D) (moving 3D images). Most pregnant women in developed and many in developing countries undergo at least one obstetric ultrasound scan during their pregnancy, and some have many more. Such scans are traditionally used to monitor the growth and development of the unborn body and to check for any malformations, but since the turn of this century the phenomenon of 3D/4D ultrasound used purely for 'social' or 'bonding' purposes has emerged.

Before these technologies were invented and attracted widespread use, with the exception of women who had experienced a miscarriage, few people from outside the medical sphere would have set eyes upon unborn entities in their earlier stages of development. The use of preserved embryo and foetus specimens for teaching purposes in medical schools became standard practice in many developed countries by the end of the nineteenth century (Dubow 2011, Morgan 2009), but these were not available for viewing by members of the public. From the early years of the twentieth century, visual displays of the inside of the human body began to be shown in fairs and exhibitions (Erikson 2007). Preserved embryos and foetuses were displayed in some museums and at state fair side shows and large scientific fairs such as the 1933 World's Fair in Chicago. Photographs of embryo and foetal specimens were also published in popular magazines and books during this period (Dubow 2011). However the vast majority of the lay population still did not have access to such artefacts.

The focus on unborn entities has also been stimulated over the past three decades or so by the introduction of the use of human embryos in assisted conception methods, scientific research and regenerative medicine. The introduction in the late 1970s of IVF techniques, in which embryos are created in the laboratory and then implanted into the uterus to achieve pregnancy, resulted in popular attention being drawn to the figure of the embryo. So too, the emergence 20 years later of the new field of regenerative medicine has aroused a whole new set of debates about the use of surplus IVF embryos for research or medical purposes. Regenerative medicine involves the attempt to replace or regenerate

DOI: 10.1057/9781137310729

diseased or malfunctioning human cells, tissues or organs, particularly those affected by chronic illnesses such as diabetes and cardiovascular disease and injuries or degenerative conditions affecting the central nervous system such as multiple sclerosis, paralysis and cerebral palsy. Various human stem cells (cells that can differentiate into diverse specialised cell types) have been employed in research attempting to assess the viability of this method of regeneration, including cells taken from umbilical cord blood. However human embryonic stem cells (hESCs) are considered the most promising because they are pluripotent, or have the ability to generate any cell in the human body, and also because they have the ability to replicate themselves indefinitely. Stem cell lines produced from these cells have been used for research and medical purposes. Isolating the inner cell mass of the early embryo (the blastocyst) from which the stem cells are derived destroys the embryo, thus generating disputes over the meaning and moral value of both the embryo and this form of medical science.

Another kind of embryo that has been configured due to new laboratory techniques is the organism that is sometimes referred to as a 'non-embryo' or an 'embryonic entity'. This is an organism that is made from techniques such as cloning or induced parthenogenesis and therefore does not involve the uniting of a male and female gamete or a unique genetic blueprint. As such, this 'non-embryo' offers scientists the opportunity to avoid the mire of ethical and moral issues around the use of other types of embryos. The 'non-embryo' has been approved for use in hESC research in the UK, Sweden and Israel (Metzler 2007). The induced pluripotent stem cell (iPSC) has also recently been investigated as an alternative to hESCs. This is an adult stem cell that has been artificially altered into a pluripotent stem cell similar to hESCs, thus offering the potential of avoiding the use of stem cells derived from embryos. The efficacy and use of iPSCs are still very much in the experimental stage, however, and it has not yet been determined whether these cells can successfully replace hESCs (Jha 2011).

As a result of all these changes and developments, unborn entities have become dominant figures not only in medical and scientific contexts but also in popular culture. Embryos and foetuses receive intense public attention, with images derived from ultrasound, foetal photography and computerised visualisations constantly appearing in popular culture. Websites and social media platforms have facilitated the wide dissemination of unborn images in their various stages of development from

DOI: 10.1057/9781137310729

conception to birth. Professional companies selling services such as IVF and 3/4D ultrasounds, websites offering information to pregnant women and those produced by anti-abortion organisations as well as individuals or couples wanting to share ultrasound pictures and videos of their unborn with others via digital media have contributed to the proliferation of the 'public foetus' and now, increasingly, the 'public embryo'.

One of the main contentions of this book is that the terminology that is adopted to refer to the products of human conception is inevitably politically, culturally and emotionally charged. When one refers to the 'unborn baby' rather than the 'foetus', for example, one is already suggesting a stance that positions this entity as already an infant, already a person. The more technical medical term 'foetus' may appear less emotive, but again its usage may represent a certain political position (for example one that refutes the concept of this entity as a person/infant). Throughout the book I have attempted to use terminology that is as neutral as possible given the constraints of the available language. While I do also often refer more specifically to 'embryos' and 'foetuses' throughout this book where appropriate, I have chosen to use the term 'unborn' more generally to denote any type of organism produced from the union of human gametes, whether *in vivo* (created in the female human body) or *ex vivo* (created in the laboratory), whether it is destined to become an infant or not. Using the term 'unborn' may suggest that this organism is destined for birth eventually. However in many cases, such as in elective abortion or spontaneous pregnancy loss and also in relation to the hundreds of thousands of *ex vivo* embryos worldwide that are produced for IVF purposes but which may be surplus to requirements or considered of not high enough quality for uterine implantation, embryos and foetuses do not survive to reach birth and become infants.

In the medical literature the term 'embryo' is used broadly to denote the conceptus (product of fertilisation of a human ovum by a human sperm) from the moment of fertilisation until the eighth week of development (tenth week of gestation), after which point it is termed the 'foetus' until birth, when it then becomes in medical terminology an 'infant', or in lay terms a 'baby'. A more specific definition that is also used in the medical or scientific literature technically defines the conceptus as an 'embryo' from around the fifteenth day after fertilisation, beginning from the formation of the 'primitive streak' from which develops all subsequent embryonic and foetal tissue. Before the fifteenth day of development, the conceptus has sometimes been titled the 'pre-embryo' in the medical

DOI: 10.1057/9781137310729

literature. The term 'embryo' also encompasses the 'zygote' (the conceptus following fertilisation until about day four or five of development, when the cells begin to divide) and the blastocyst (the conceptus from about day 5 to day 14 of development). For the sake of simplicity, the discussion in this book will generally use the term 'embryo' to encompass all of these developmental stages unless greater specificity is required.

These very precise scientific definitions fail to recognise the blurring of the boundaries between embryo, foetus and infant that regularly takes place both in medical and popular discourses and images. This ambiguity and the sociocultural meanings, practices, discourses and affective responses underpinning it is one of the central concerns of this book. The book's title refers to the 'social worlds' of the unborn. A large literature on the sociology of the body has investigated the ways in which human bodies are configured and experienced via the interaction of flesh, others' bodies, discourse, practice, material objects, space and place. I contend that this approach to embodiment may be extended even earlier, to unborn bodies, and it is this perspective that is adopted in the present discussion. Unborn bodies may be understood as constantly changing configurations produced by their interaction with a range of heterogeneous elements, human and non-human, ideational and material. I take the theoretically relativist position which argues that from the time a conceptus is created, and indeed even before, when it is a 'preconceived embryo' (Karpin 2010), its meaning is shaped via cultural and social understandings. The unborn from this perspective are complex entities that are composed of medical and scientific practices, technologies and physical spaces, but also of social relations, interpretations and understandings between human actors that are part of a constantly negotiated social order.

As I seek to demonstrate throughout this book, how we think about, treat, represent and monitor the unborn has changed over and across historical periods, cultures and geographical locations, and remains open to change. The social values and meanings we ascribe to embryos and foetuses are inevitably shaped via their location in these contexts. As Dubow (2011: 3) has put it, 'A fetus in 1970 is not the same thing as a fetus in 1930, which is not the same thing as a fetus in 1970, which is not the same thing as a fetus in 2010'. Embryos and foetuses may therefore be considered as 'social objects' (Casper 1998, Ehrich *et al.* 2010, Franklin 2006a, 2006b, Morgan 1996, 2009, Taylor 1992). They may further be viewed as 'boundary objects' (Star and Grisemer 1989), or objects that

DOI: 10.1057/9781137310729

move between related but different social worlds. Boundary objects often have different sociocultural meanings in the worlds they transverse. However, these meanings may change and be 'translated' readily from one social world to another.

Some meanings are shared between social worlds, through shared scientific processes of work, for example. The work practices and technologies employed by those professionals who deal with embryos or foetuses *in utero* and who make and handle *ex vivo* embryos are central to the construction of these meanings. These practices and technologies differ according to the ways in which the unborn are categorised, the purposes for which they are created and assessments of their quality and appropriateness for the various uses to which they can be put (Casper and Morrison 2010, Ehrich *et al.* 2006, Wainwright *et al.* 2006, Williams *et al.* 2008).

The social worlds across which the unborn move may now also be located in different geographical areas: across individual countries or the globe. Recent years have witnessed the growth in the phenomenon of 'fertility' or 'reproductive tourism', in which individuals or couples travel to other countries to engage the services of IVF clinics, gamete donors and pregnancy surrogates because these services are less expensive, there is greater access to donated gametes or the legislation concerning these services is more lenient in those countries. As a result, embryos and foetuses are becoming part of a transnational economy in which they may be created via IVF in one location, sometimes using donated ova or sperm that have been cryogenically preserved and then transported from another location, and then may be moved to yet another location to be gestated within the body of a pregnancy surrogate.

The term 'assemblage' is now often used to describe the diverse and constantly changing configurations of human bodies/selves (among other phenomena). This concept emerged from the philosophy of Deleuze and Guattari in conjunction with actor network theory, a major theoretical approach in science and technology studies (Marcus 2006). The concept of the assemblage acknowledges the constantly shifting and contingent nature of embodiment and subjectivity and the importance of recognising the interaction of bodies with others' bodies and with non-human agents. It highlights the components from which phenomena – including human bodies and subjects – are comprised and with which they relate. The concept of the assemblage goes beyond a social constructionist perspective, which tends to assume that once a phenomenon has been

DOI: 10.1057/9781137310729

constructed its meanings are fixed, to a perspective which allows for constant change and the interplay of meaning between social actors in the making and remaking of phenomena (Casper and Morrison 2010).

Given the integral role played by ideas, discourses, practices and material objects such as medical and other scientific technologies in making, visualising, representing, measuring and monitoring unborn entities, I would contend that such a perspective offers much to the understanding of the social worlds in and across which the unborn move and are configured. I use the terms 'unborn assemblage', 'maternal assemblage' and the hybrid 'unborn–maternal assemblage' in this book to denote the interrelationship between the unborn and the pregnant women who harbour many (but importantly, not all) of the unborn within their own bodies. There is no single 'unborn assemblage'. Just as any other human body is open to change and contestation in its meaning, unborn assemblages are mutable, changing in form not only as they physically grow and develop but also as the social worlds in which they are located shift and change.

This book examines changes in configurations of unborn assemblages across the multiple social worlds they inhabit. It includes discussion of notions of the unborn prior to the development of such devices as obstetric ultrasound and how the concept of the hidden, mysterious organism that was inextricably part of the maternal body changed to encompass that of the unborn entity as a miniature individuated human subject with its own rights and privileges. While I focus predominantly on developed societies, there is also some examination of how the unborn are conceptualised and dealt with in developing societies, serving to emphasise the contingent nature of such concepts and practices. As the book will demonstrate, even within the developed world there are significant differences between countries in the ways in which such phenomena as elective abortion, IVF and hESC research are viewed and regulated, based on social, cultural, political, historical and religious determinants that work to shape national perspectives, opinions and legislation.

Feminist critics have for some time directed attention to how concepts of pregnant women's bodies have intersected with those of the unborn and how the rights of each entity have been conceptualised in relation to the other. Such topics as contraception, abortion, assisted reproductive technologies, sperm and ova donation, surrogate pregnancy and prenatal screening technologies provoked many interesting and challenging

DOI: 10.1057/9781137310729

feminist analyses from the 1980s into the 1990s (see, for example, Bordo 1993, Cohen 1996, Franklin 1997, Ginsburg 1990, Haraway 1991, Hartouni 1991, 1992, Hubbard 1984, Martin 1992, Oakley 1984, Petchesky 1980, 1987, Rapp 1990, Rowland 1992, Warren 1988). Since the turn of the twenty-first century, new developments in reproductive and prenatal screening and visualising technologies, the growing market in reproductive tourism and the use of embryos for regenerative medicine have inspired a renewed flurry of commentary from feminist and other scholars on the moral, ethical, cultural, social and political dimensions of the unborn, many of which will be referred to in the pages of this book.

Given the contentious nature of the discourses and practices around the *ex vivo* embryo, it is not surprising that most of the recent scholarship in the humanities and social sciences has directed attention at this specific type of unborn entity. Yet it is the *in vivo* entity with which most people in everyday life are familiar and have personal experience, particularly at the foetal state of development. There are a multitude of popular cultural representations of the unborn that do not specifically refer to or which predate IVF or hESC science. Unborn entities have a longer history and a more diverse constellation of meanings that contribute to and also draw from contemporary science- and technology-related representations. Women's embodied experiences of conceiving, gestating, miscarrying or aborting embryos and foetuses or producing and using them in IVF treatment, lay understandings of the unborn, political debates and portrayals of the unborn in the popular media intertwine and interrelate with each other and with medical and scientific ideas, practices and technologies to configure and reconfigure unborn assemblages.

This book examines all of these topics and issues, drawing upon relevant literature from sociology, anthropology, media and cultural studies, bioethics, science and technology studies, social history, cultural geography, philosophy and gender studies to do so. I identify and examine the different, but often overlapping, social worlds in which the unborn are configured and in and across which they move: the home, medical and obstetric ultrasound clinics, the laboratory, the mass media, online digital platforms, commodity culture, legislative and political arenas and so on.

Chapter 1 reviews some social, cultural, historical and ethical approaches to unborn entities, focusing on debates in the literature concerning the status of the unborn as persons and as human. It includes discussion of concepts of the unborn in previous eras in developed societies as well

DOI: 10.1057/9781137310729

as in contemporary developing societies to highlight the contingency of meanings around embryos and foetuses. The role of definitions in shaping concepts of and practices around the unborn is also covered in this chapter. Chapter 2 details the various ways in which visualising devices have been used to depict and represent the unborn body, including photographic and ultrasound images, computerised visualisation technologies and portrayals in the news and social media. It addresses the commodification of the unborn image and examines how imagery has been used for political purposes. In Chapter 3 I examine the ontology of the unborn–maternal assemblage from the perspective of pregnant women. The discussion draws both on empirical research into women's experiences of pregnancy and feminist philosophical analyses of pregnant embodiment. Chapter 4 goes on to address the topics of abortion, pregnancy loss and the disposal of surplus IVF embryos. While these are very different issues, they all relate to the ways in which the unborn are understood in relation to the continuum of 'life' or 'personhood'. In the final substantive chapter, Chapter 5, I look at the intensification of discourses and practices that position the unborn assemblage as both precious and vulnerable and as endangered by the actions (and even emotions) of the maternal body in which they develop. Here I identify how the unborn assemblage is configured in these discourses as both separate from and joined to the maternal assemblage. In the brief Final Thoughts section I bring the arguments of the book together and make some suggestions for alternative ways of thinking about unborn–maternal assemblages.

DOI: 10.1057/9781137310729

1
Contingencies of the Unborn

Abstract: *Should the unborn be viewed as fully 'persons' in moral and ethical terms? Should they be considered indeed fully 'human'? What are the implications for how they are treated in medical and scientific procedures, and for the pregnant women in whose bodies many, but not all, unborn entities are located? As this chapter will demonstrate, the answers to all of these questions are arbitrary, located within specific social, cultural and historical contexts. It includes discussion of concepts of the unborn in previous eras in developed societies as well as in contemporary developing societies to highlight the contingency of meanings around embryos and foetuses. The role of definitions in shaping concepts of and practices around the unborn is also covered.*

Key words: sociology; anthropology; history; embryos; foetuses; personhood; ethics

Lupton, Deborah. *The Social Worlds of the Unborn.* Basingstoke: Palgrave Macmillan, 2013. DOI: 10.1057/9781137310729.

DOI: 10.1057/9781137310729

Personhood, humanness and the unborn

The contested nature of defining the unborn came to the fore when a highly controversial article published in 2012 in the prestigious international *Journal of Medical Ethics* put the case for the logic of 'after-birth abortion' (Giubilini and Minerva 2012). The authors, two bioethicists, Italian Alberto Giubilini and Francesca Minerva from Australia, assert in the article that the killing of a newborn infant, regardless of its state of health or normality, should be viewed as ethically acceptable if its birth were seen to disadvantage its parents or siblings in some way, including imposing a psychological or economic burden upon them.

Giubilini and Minerva use the terms 'pre-birth abortion' and 'after-birth' abortion to suggest that there is little difference between the foetus and the newborn infant. To support their argument, Giubilini and Minerva claim that neither the foetus nor the newborn infant 'can be considered a "person" in a morally relevant sense' (2012: 2) because they lack the properties that justify the attribution of a moral right to life. They argue that both should be considered 'potential persons' rather than fully developed persons. Because of their undeveloped nervous system and lack of awareness, neither foetus nor neonate can be said to possess values or beliefs. Neither is capable of attributing value to life, a perspective on its own interests or forming long-term aims, as they have not achieved the requisite mental development. Foetuses and newborn infants are therefore not capable of recognising any harm to themselves from being killed. Based on this reasoning, Giubilini and Minerva claim if 'pre-birth' abortion were considered morally and legally allowable, as it is in many jurisdictions (albeit usually with restrictions concerning gestational age – see Chapter 4), then so should 'after-birth abortion'.

Such logical and pragmatic utilitarian arguments fail to acknowledge the sociocultural and emotional meanings that gather around and constitute the unborn. These meanings came to the fore, however, when Giubilini and Minerva's article attracted public attention immediately following its publication, when it received a high level of coverage in newspapers and websites. The authors' arguments aroused major debates over their article's moral worth. They were subjected to death threats and the journal's editors were also sent abusive and threatening messages. The journal's website was deluged with responses. Comments on various websites commenting on the article, including the journal's own, referred to the 'evil', 'vile' and 'scary' nature of the arguments posited in

DOI: 10.1057/9781137310729

the article and described the act the authors were advocating as 'murder' and an 'atrocity'. Giubilini and Minerva were even described as 'soulless monsters' in one comment on the journal's website.

In response to the criticism, the editor, the well-known Australian applied ethicist Julian Savulescu, was forced to publish a blog entry on the journal's website defending his decision to publish the article (2012). Savulescu is himself known for his utilitarian approach to prenatal genetic diagnosis and screening and his support of hESC research. In his defence of Giubilini and Minerva, Savulescu pointed out that infanticide has been the subject of many scholarly ethical discussions, including in his own journal, and that there has been an active debate over the moral status of the unborn and neonates in philosophy for many years. Savulescu contended that Giubilini and Minerva were simply the latest in a long line of philosophers raising questions about the extent to which the unborn and the newborn should be considered fully 'persons'.

One commentary published on the journal's website in response to Giubilini and Minerva's article was penned by Charles Carmosy, a Roman Catholic theologian. The title of his piece referred to 'Our vulnerable prenatal and antenatal children' (2012). This terminology of the 'prenatal' and 'antenatal' child flags the orthodox Catholic view, promulgated by the Vatican, that unborn humans are already children. As a catechism of the Catholic Church puts it, 'From the first moment of his [sic] existence, a human being must be recognized as having the rights of a person – among which is the inviolable right of every innocent being to life' (*Catechism of the Catholic Church: Part Three – Life in Christ* 2013). Carmosy points to the relational nature of the unborn human and argues that simply because the potential of that human is immature or frustrated (as in the case of adults with mental disabilities, for example), this is not a just reason to consider them less than fully persons. He also highlights the difficulty of deciding exactly when a newborn might be considered to have developed the attributes of what Giubilini and Minerva consider 'actual personhood'. For Carmosy, the very vulnerability of the unborn is what gives them the 'special' status of requiring greater rather than less protection, underpinned, he argues, by the Christian ethic of protecting the vulnerable.

What the arguments both of bioethicists such as Giubilini and Minerva and of theologians such as Carmosy attempt to do is to conflate the unborn and the already born so that they are seen as morally and ethically the same. For Giubilini and Minerva both foetuses and

DOI: 10.1057/9781137310729

neonates are lacking in full personhood; for Carmosy, both are fully human children deserving of the kind of protection and rights offered to any other human.

As this debate and the public response to it highlights, agreement concerning at which point in human development the unborn organism is considered to be a 'human' or fully a 'person' has long been subject to contestation and has varied across western history and across cultures (Casper 1994a, 1994b, Conklin and Morgan 1996, Duden 1993, Featherstone 2008, James 2000, Kaufman and Morgan 2005, Morgan 1996, 1997, 2009). Definitions of personhood, and by extension, notions of individual embodiment, are constructed via social and cultural understandings. They are dynamic and shifting, open to change and contestation. What is considered 'human', and in contrast, 'non-human', are themselves contested and constructed subject positions rather than fixed or natural, subject to claim and counter-claim. 'Doing human' is performative and recurring, based on practices and interactions (Casper 1994a, 1994b). As Conklin and Morgan argue, '[t]he beginning of life – the time when new flesh must be interpreted, shaped and transformed into socially meaningful forms – is especially revealing of how competing views of personhood are "worked through the body"' (1996: 663). Personhood, therefore, is both biological and social, both natural and cultural, phrased in different ways according to the specific context in which it is debated and understood (Cesarino and Luna 2011, Marsland and Prince 2012).

Indeed discussion of the contingencies of the unborn go directly to the heart of contemporary scholarship on 'vitality' and how we define it, including concepts of when life begins or ends (Kaufman and Morgan 2005, Marsland and Prince 2012). Building on the work of Foucault on biopower and biopolitics, several writers (Rose 2007a, 2007b, 2009, Waldby 2002a, 2002b, 2008, Waldby and Squier 2003) have explored the question of human 'life itself': how vitality is understood, how it is valued, the strategies, discourses and practices that configure it and how boundaries are constantly blurred between 'human' and 'non-human', 'living' and 'inanimate', 'material object' and 'flesh' and so on. Such inquiries have highlighted the role played by the concepts, discourses and practices of scientific medicine and other forms of biotechnologies in defining the human body and its capabilities and ills. Medicine and biotechnologies have played an increasingly important role in the configuration of the unborn, including producing new forms of life, such

DOI: 10.1057/9781137310729

as *ex vivo* embryos and the stem cell lines that are generated from them. Whether or not an unborn entity is created *in vivo* or *ex vivo*, its trajectory of development is observed, monitored and regulated by medical and other scientific practices, including the forming of decisions about how valuable, viable, normal and capable (or worthy) of life it is.

There is something unearthly, strange and Other about the unborn entity. Given its early stages of development, its temporal distance and morphological difference from the newborn infant, the embryo is a particularly ambiguous, complex, hard-to-pin-down entity. At the embryonic stages of development in particular, the unborn body appears like another creature gradually morphing into a human-like body. It is not until about the tenth week of gestation (eight weeks following fertilisation), at which point in development the embryo technically becomes a foetus, that it begins to look human. Even at the foetal stage of development, the unborn body still has a certain strangeness to it, a liminality that may challenge accepted concepts of humanness and of living creatures. As McGinn asserts:

> The fetus is not yet an autonomous living being, more like a bloated internal organ than a fully functioning creature, quite unable to survive outside the mother's body. It is an almost-life, at risk of extinction, poised between nothingness and existence, an intermediate identity: in it we see a precarious quasi-life, a fragile upsurge from the emptiness that precedes it. (2011: 100)

This quotation comes from a book-length analysis of disgust, in which the author puts forward the proposition that entities that are not easy to categorise, that inhabit the border between life and death, human and non-human, provoke disquiet and even feelings of disgust because of their liminal status. While the corpse is the apotheosis of a bodily object of disgust, unborn entities (and even newborn infants) may similarly arouse feelings of unease and revulsion because they do not easily fit into these cultural categories. McGinn goes so far as to suggest that 'As the corpse is a repulsive token of life's ambiguous end, so the foetus and the new-born are semi-repulsive tokens of life's fraught beginning' (2011: 101).

In direct contrast to this notion of unborn entities as somewhat repellent and unsettling and not quite (or yet) human, however, is the increasingly common portrayal of them as already loveable and cute 'babies'. This ontological extension backwards of infancy in the human developmental time-frame has major implications for how we think about and treat the unborn. The growing emphasis in wealthy developed societies

DOI: 10.1057/9781137310729

on the value of children and the accompanying discourses of intensive parenting (Beck and Beck-Gernsheim 1995, Hays 1996) has resulted in children being considered infinitely important (Ruddick 2007, Zelizer 1985). This is particularly the case for infants, considered the most vulnerable and precious of all children (Lupton 2013a, 2013b). I have commented elsewhere (Lupton 2013b) on the four dominant discourses that frame notions of the infant body in the popular media: as precious, pure, vulnerable and uncivilised. These discourses are also all relevant to concepts of the unborn. Both the unborn and infants are considered highly valuable, the most important of all humans in terms of their potentiality, their closeness to nature and their purity, uncontaminated as they are by the excesses of culture. Their bodies are viewed as highly fragile and open to contamination by polluting outside influences, and therefore as requiring high levels of protection. Indeed the unborn have increasingly become positioned as even more vulnerable to such threats than are infants, given that they are viewed as at the mercy of their mothers' actions, unable to escape the confines of the womb (see more on this in Chapter 5). As I will go on demonstrate in Chapter 3, the unborn may also be conceptualised as 'uncivilised' by the pregnant women in whose bodies they are developing if they are viewed as antagonists, creating discomfort or illness for women.

The *ex vivo* embryos that have become such a focal point of public discourse and policy-making in the past 15 years or so in relation to their disposal and their use in hESC research and regenerative medicine are at the blastocyst stage of development, and thus are only about five days past fertilisation. They have no human appearance at all: they are simply a round-shaped collection of dividing and multiplying cells. Yet for those couples who have undergone IVF, the opportunity to view their embryos at this very early developmental stage through an electron microscope can be an important way in which they come to visualise and think about these products of conception. For some people this process of viewing the blastocysts, even though at this stage they have no human form, serves to reinforce the notion of these entities as the beginning of life, the precursor to their 'baby' (Haimes *et al.* 2008, Kato and Sleeboom-Faulkner 2011). As one Japanese woman who was undergoing IVF commented, 'When I saw my embryo through the microscope, I thought that I had finally met my child' (Kato and Sleeboom-Faulkner 2011: 439). However this is not a universal reaction. For other people undergoing IVF, viewing their embryos supports their concept of these

DOI: 10.1057/9781137310729

organisms as 'just cells' that have not attained personhood (Haimes *et al.* 2008: 121). In an Australian study, for example, one man described the appearance of the blastocyst he had viewed as 'just a little blob', likening its appearance to 'bubbly eggwhite', while a woman commented that her embryos 'don't look real' (as foetuses or infants do) (de Lacey 2007).

As this suggests, uncertainties about whether the unborn are fully human and fully persons, and at what point in their development they achieve these statuses, begins at the very earliest stage of embryonic development. The sociocultural, historical and political contexts in which the unborn are encountered shape concepts of their personhood and which stage they are viewed as reaching in the journey to full human subjectivity. The unborn, particularly in the earliest days and weeks of their development, are portrayed differently across various lay, medical and scientific texts and practices, as human and non-human, alive or not yet alive according to the context. They are placed on the margins of the boundaries between human and non-human or in the spaces between these categories. They are thus located on a continuum rather than a binary definition of humanness (Casper 1994a, 1994b, 1998, Cesarino and Luna 2011, Christoffersen-Deb 2012, Conklin and Morgan 1996, Haimes *et al.* 2008, Morgan 2009).

Anthropologists have demonstrated that different cultures have differing ways of deciding when unborn and even post-born entities are deemed to become 'infants' or even designated as 'human' (Conklin and Morgan 1996, Gottleib 2000, James 2000, Kaufman and Morgan 2005, Littlewood 1999, Morgan 1997, 2006a, 2009). There are 'processes of coming-into-social being' (Kaufman and Morgan 2005: 321) which are related not to biological attributes but rather to accepted understandings within a social group or culture. James (2000) draws on the writings of the early anthropologist Marcel Mauss to argue that that these processes involve a 'recognition' or pragmatic acceptance of an entity as a separate being from the maternal subject and at least provisionally a 'person'. This is not universal or automatic for unborn or even already born entities. As James notes, these processes of recognition are expressed with wide variation between cultures, but they centre on the concept of individuation from the mother and identification of a social being that can respond to others and demonstrate a potential to engage in reciprocal social relations.

Some cultures locate the beginning of infancy while the unborn body is still *in utero*. As I argued earlier and will throughout this book, this extension of infancy back into gestation has become increasingly apparent in some countries and cultures. In many other cultures, however, even newborn infants are considered to be not fully human until they have demonstrated certain behaviours or lived for a defined time period (Conklin and Morgan 1996). In one Australian Aboriginal culture, for example, the newborn is still called a foetus for some weeks until it has started to smile, and only then is it called a child (Gottleib 2000). In Ecuador the foetus remains a liminal organism, a creature rather than a human until birth and beyond, with infants not being accorded full personhood until some months after birth. As this research demonstrates, some cultures do not make a definite, clear-cut distinction between 'person' and 'non-person' when thinking about the unborn: they incorporate the category of 'quasi-person' to describe humans who are considered to occupy an ambiguous status. In the case of the Ecuadorians, this status of quasi-person is attributed to humans who have not been baptised according to their religious faith, such as infants who have not yet undergone the ceremony and the unborn (as well as adults who have not been baptised) (Morgan 1997, 2009). The unborn arouse even more uncertainty about personhood and individuated embodiment than do newborns, given that they still are encased within the maternal body.

Histories of the unborn

It is important to emphasise that concepts of the unborn in many societies have changed considerably in recent times. Historical analyses, like cross-cultural comparisons, are able to demonstrate the contingent nature of the unborn. They have shown that in previous historical eras in western societies very different notions of the unborn circulated. Despite their potent contemporary position as cultural icons, the unborn as they are represented and conceptualised today are relatively recent figures.

The concepts and representations of the unborn produced through medical and scientific knowledge emerging in the early modern period in Europe, for example, portrayed these entities in very different ways from previous representations. Before the invention of medical and scientific technologies that could conduct anatomical dissection or autopsies on unborn corpses, test for pregnancy hormones, visualise the unborn body or hear its heart-beat, the unborn, particularly in their earlier stages of development, were enigmatic, hidden creatures. One major

DOI: 10.1057/9781137310729

difference between the contemporary era and earlier times – as recently as a generation ago – is that embryos and foetuses were never referred to using these medical words (Duden 1993). While medical drawings and anatomical wax figures from the early modern period represented the foetal body as individuated from that of the maternal body, lay people were not exposed to these representations and did not think of the foetus in these terms (Duden 1993, Erikson 2007, Hanson 2004, Newman 1996). Indeed the unborn body was not routinely treated legally or clinically as if it were a separate entity until the 1960s (Featherstone 2008, Weir 1998). The unborn–maternal assemblage was inextricably interbound and considered as a unitary organism until the moment that the unborn passed out of the maternal body, at which point they became viewed and treated as separate entities.

Duden (1993) argues that in the pre-modern era people described to physicians the haptic sensations, or feelings derived from the senses of touch, smell and taste and perception of space and atmosphere they experienced inside or with their bodies, rather than the physician examining them and relying upon what he could see to make a diagnosis. For women in the first months of pregnancy the unborn body was experienced as an invisible constellation of bodily sensations and physical symptoms only they could experience, and others relied upon the women to recount these. It was the pregnant woman, therefore, who bore ultimate authority and knowledge about her unborn. The concept of 'quickening', in particular, was used to describe the first movements of the foetal body felt by the pregnant woman and provided confirmation that she was pregnant. As only the pregnant woman could feel these movements at first, during the earlier months of pregnancy it was her word only upon which judgements of pregnancy could be made. Even in later pregnancy, the pregnant woman retained her unique knowledge of the unborn gestating within her. Other people could only view the woman's swelling abdomen or see or feel some of the movements of the foetal body within her (Duden 1993, Featherstone 2008).

Thus, before the advent of scientific medicine the unborn may have given signs and intimations of their presence, but there was never full certainty as to their existence until they actually passed out of the pregnant woman's body. Women could therefore not be legally prosecuted for procuring an abortion (Duden 1993, Featherstone 2008, Hanson 2004). Nor was a miscarriage viewed as the death of a human or even a human-like entity during this era. As recent as the late eighteenth century in

DOI: 10.1057/9781137310729

Germany, for example, women and their physicians used such terms as 'blood curds', 'wrong growths', 'inconsistent beings' and 'fleshy morsels' to describe the products of women's uteruses when a miscarriage had occurred (Duden 1993: 64–65).

The period spanning the late eighteenth and the early twentieth centuries was marked by the rapid growth of scientific and medical knowledge about the human body and thus was a pivotal era for transformations in ways of viewing, treating and monitoring the unborn. Once scientific medicine had established in the late nineteenth century that fertilisation involved the meeting and union of ovum and sperm, and had developed greater knowledge about embryonic and foetal development from autopsies and vivisection, the unborn became increasingly viewed as human and alive from an earlier stage (Dubow 2011, Featherstone 2008). The invention of the stethoscope, which allowed the foetal heart-beat to be heard for the first time, also contributed to new ways of 'knowing' the unborn and monitoring its presence and wellbeing while it was still within the maternal body (Hanson 2004).

The development of the specialist medical field of embryology at the end of the nineteenth century, and gaining momentum into the early decades of the twentieth, resulted in a vast increase in medical knowledge about the earliest stages of human embryonic development. The pioneers in this field in Europe and the USA collected tens of thousands of embryos from miscarriages, gynaecological surgical procedures, ectopic pregnancies, autopsies and abortions to use in their research and dissected them to develop their knowledge of embryonic morphology and stages of development (Morgan 2002, 2006b, 2006c, 2009). Embryonic developmental forms were analysed, described and visually represented – first in anatomical drawings, then in photographs – in minute detail (Morgan 2009).

The findings of embryology and better understanding in obstetrics of unborn development resulted in changes to the ontological and moral position of the unborn. Once the figure of the embryo had been documented and visualised in anatomical drawings, early abortion before 'quickening' became viewed in legal circles as a criminal act rather than as the removal of 'troublesome blockages', because the organism was now defined as a potential child. The embryo was still not legally considered a separate person in its own right, however (Duden 1993). By the late nineteenth century, doctors began advising pregnant women to protect their unborn's health by engaging in certain protective behaviours. A growing

DOI: 10.1057/9781137310729

sense of the foetus as an individual began to develop. This was expressed in attitudes towards caesarean section. At that time this procedure was relatively safe for the foetus but extremely dangerous for the mother, but yet was advocated to protect the life of the foetus (Featherstone 2008).

The rapidly developing knowledge of embryonic development contributed to the generation of a movement within the USA on the part of physicians seeking to proscribe induced abortions on the basis that conception, rather than foetal quickening, was the beginning of independent unborn life. Part of their argument was that the unborn and the pregnant women in which they were developing should be represented as entirely distinct from each other, as separate lives (Dubow 2011). By the early twentieth century in the USA, the unborn had been identified as separate legally from the maternal body to the extent that abortion had been criminalised in almost every state (Dubow 2011).

The emergence of visual imaging of embryos and foetuses in the 1960s using photographic techniques and obstetric ultrasound provided the starting point for images of the unborn that could be readily accessed by the lay population. It is from this time onwards that embryos and foetuses became public figures, their images frequently reproduced in popular magazines, coffee table books and television documentaries. The detail offered by the beautiful images of the unborn captured by the pioneering Swedish photojournalist Lennart Nilsson allowed members of the public to observe *in utero* life in close-up from the moment of conception to birth (see Chapter 2 for more discussion of Nilsson's work). The real-time ultrasound technology capable of capturing foetal movements rather than static images was developed in the mid 1960s (Erikson 2007). Once obstetric ultrasound become a commonly used technology in the late 1970s, the once opaque and secret environment of the uterus, perceptible only to the pregnant woman herself via her physical sensations, was opened to public observation. The unborn body could now be seen *in utero* moving about, with recognisable features and limbs. As a result, haptic perceptions of the unborn became largely supplanted by optic or visual interpretations. The pregnant woman was no longer the ultimate authority on her unborn, for other forms of configuring the unborn assemblage that required the technologies and expert knowledge of medical science had now gained credence (Duden 1993).

The introduction of paediatrics as a new medical speciality at that time contributed to a medical view that foetal health and wellbeing should be separated from maternal health and wellbeing. The foetus became

DOI: 10.1057/9781137310729

viewed and treated as a patient that, while still part of the maternal body, could also be individuated from the mother in some ways. This change heralded the beginning of a reconceptualisation of the foetus as infant in medicine (Featherstone 2008). Until this time, obstetrics focused very little on the foetus as an entity requiring medical intervention or care, directing its attention predominantly on birth and on the health and wellbeing of the pregnant and labouring woman (Hanson 2004, Weir 1998). Developments in obstetric ultrasound allowing the production of finer-grained imagery in the 1970s led to new medical specialties: foetal cardiography, foetal biometry (the measuring of foetal body parts) and prenatal testing techniques such as amniocentesis and chorionic villus sampling (Erikson 2007). The emergence of foetal surgery, in which foetuses are operated upon while *in utero*, is another medical technology that grew from the diagnostic possibilities offered by ultrasound. It presented yet another way in which to separate the bodies of pregnant women and their foetuses and represent the foetus as an individual (Casper 1994a, 1994b, 1998, Ruddick 2007, Van der Ploeg 2004, Williams 2005, 2006). Intrauterine transfusions developed as a technique in the late 1960s and 1970s, heralding the advent of a new medical speciality: foetal medicine, or foetology (Dubow 2011).

Once a foetus becomes designated as a 'patient', the maternal body in which it is developing tends to become viewed as an 'environment' or even an 'incubator' for the nourishment and protection of the unborn. The implications of treatment of the foetus for the pregnant woman's health and wellbeing may be glossed over or ignored when the foetus is the primary focus of medical attention (Casper 1998). Thus, for example, medical writings about experimental foetal surgery in the 1990s portrayed the foetus as a living human being, an embodied actor within the pregnant woman's uterus. Practitioners working in the field routinely referred to the foetus as 'the kid' or 'the baby', while in some medical writings the pregnant woman was represented as the 'organic host' for the foetus (Casper 1994a, Williams 2005). Medical journals also routinely made reference to surgical procedures being carried out on the foetal body without mentioning the maternal body which was also being operated upon. They thus removed the maternal body from their purview altogether. In instances where the foetus's health deteriorated or it died following a surgical procedure, the 'mother' was said not to have suffered any complications or morbidity from this surgery, for the foetus was not viewed as part of her and she was not positioned as the

DOI: 10.1057/9781137310729

patient. The pregnancy was described in these texts as happening to the foetus and not to the woman (Van der Ploeg 2001, 2004). Medical texts also commonly represented the prenatal period as conterminous with the postnatal period, representing pregnant women as already 'mothers' with individuated infants, albeit still *in utero*. Thus the embodied and ontological distinctions between the foetus *in utero* and the newborn infant tended to be erased, further supporting the concept of the foetus as already an infant (Van der Ploeg 2001).

So too, medical research published in the 1980s directed at identifying that foetuses feel pain positioned them as patients who may require medication for pain relief or sedation if operated upon *in utero*. These ideas provided a powerful rationale for the anti-abortion movement, generating a rhetorical strategy to oppose abortion on the grounds that it subjected the unborn to suffering (Dubow 2011, Williams 2005). This argument was advanced by American President Ronald Reagan in a speech in 1984. It provoked an official rebuttal from the American College of Obstetricians and Gynecologists that there was no evidence that the unborn had developed the neurological pathways necessary to experience pain until the third trimester of pregnancy. Despite the College's statement, Reagan's speech inspired an anti-abortion activist and physician to make the notorious anti-abortion film 'The Silent Scream' (Dubow 2011) (see Chapter 2 for further discussion of this film).

The development of IVF in the 1970s was another technological initiative that wrought transformations in the ways in which the unborn were configured. The first 'test-tube' infant, Louise Brown, was born in the UK in 1978 after decades of experimentation. For the first time, unborn entities that could survive artificial processes of conception and implantation and mature into infants were successfully created in the laboratory, resulting in new ways of thinking about and treating them. Embryos could now not only be created *ex vivo* and implanted into the uterus to establish a pregnancy, they could also be cryogenically preserved (frozen) for transportation to other sites or for future use, they could be evaluated for their quality before being transferred to the uterus and the commissioning parents could view them under a microscope before implantation. The unborn also become commodities for the first time, attracting a new market value, their creation subject to monetary exchange from the couple or individuals commissioning the IVF process (or their health insurers or health system on their behalf) to the medical and technical staff responsible.

DOI: 10.1057/9781137310729

When increasing numbers of infants produced by IVF began to be born, concerns were raised from diverse quarters about the ethics, morality and health implications of intervening in the human conception and gestation process. Feminist critics, psychologists, legal scholars and ethicists expressed their concerns about the potential for women to be exploited and their reproductive powers usurped by male medical professionals, and the possible negative effects, emotional and physical, upon the women undergoing these procedures and the infants created by them (for example, Cohen 1996, Corea 1985, Lorber 1987, Rowland 1992, van Balen 1998, Warren 1988). Since 1978, IVF birth rates have progressively risen each year. It has been estimated that approximately 5 million children conceived via IVF have now been born (Hazell 2012). The sheer numbers of such births has meant that IVF embryos used for assisted conception and the children born of them have become generally accepted as 'normal', and even banal (Franklin 2006a).

Since the late twentieth century, IVF techniques and their products – in particular ova and embryos – have become the starting point for a number of scientific and medical research and therapeutic endeavours that go beyond simply assisting infertile couples to conceive. Developments in the trade of human reproductive tissue such as ova, sperm and embryos as well as the growing globalised industry in the use of surrogate pregnancy have resulted in new ways of conceptualising and treating the unborn. In these social worlds the donation of gametes, the production of embryos and the gestation of the unborn have become ever more commercialised. Their intrinsic potential in this context is that of wealth accumulation and entrepreneurialism (Ganchoff 2004, Hoeyer *et al.* 2009, Ikemoto 2009, Mitchell and Waldby 2006, Waldby 2008). In this context unborn assemblages are configured, like other human tissues such as body organs or blood plasma, as biological entities for donation or sale that are alienable from the human subject whose body provided them (Waldby 2002b).

The control and ownership of body parts in this 'tissue economy' (Mitchell and Waldby 2006) are normalised via the processes and assumptions of the biotechnology industry (Ikemoto 2009). While sperm donation in some countries has historically been remunerated, in wealthy countries ova and embryos are generally given for free – altruistically donated – or exchanged for free IVF treatment, and maternal surrogates are compensated financially only for their expenses rather than paid outright for their services. Indeed several developed countries,

DOI: 10.1057/9781137310729

including the UK and Australia, have taken steps to prohibit payment for the provision of ova and embryos and for the services of pregnancy surrogates because of concerns about the potential exploitation of donors and surrogates, just as payment for the donation of other human tissue or organs is generally banned. Some countries, such as France and Germany, have banned surrogacy – whether altruistic or commercial – altogether (Shetty 2012).

The USA is a notable exception among the developed countries: it has a flourishing market in the commodification of gamete donation and surrogacy, and pregnancy surrogates can earn up to US$30,000 for their services (Bailey 2011). In several less wealthy countries in parts of Eastern Europe (particularly Ukraine but also Russia) and Asia (predominantly India but also Malaysia and Thailand), a growing economy has also emerged, in which women are paid to produce the embryos and ova required for IVF or hESC science purposes and to act as pregnancy surrogates for wealthy westerners. Via these arrangements women effectively sell their ova or the embryos made using their ova, or in surrogate arrangements sell access to their uteruses by other people's embryos (Bailey 2011, Damelio and Sorensen 2008, Mohapatra 2012, Shetty 2012).

The unborn have therefore come a very long way from the largely unknown, mysterious entity residing in the pregnant woman's body to the organism that is frequently positioned as ontologically separate from the maternal body, every stage of its development from fertilisation onwards visualised and documented in great detail. Where once unborn entities were the property only of the women in whose uterus they resided and they had no commercial value, they may now be bought and sold, frozen and transported and grow in a different body from that to which they are genetically related. Where once the creation and development of the unborn were left to fate, numerous medical or scientific interventions may be applied to them from pre-conception onwards.

The importance of definitions

As I observed in the Introduction, language is vital in the politics of defining unborn personhood. The very use of the terms 'embryos' and 'foetus', although they are technically medically correct, is highly politicised. Their use positions the unborn as somewhat less human or adorable than do the terms 'infant', 'unborn child' and (especially) 'baby'

DOI: 10.1057/9781137310729

and such terms are chosen carefully by protagonists in contestations over such issues as abortion and hESC research. Most pregnant women do not use these technical words to describe the unborn entity growing within them, preferring to think of it and talk about it as 'my baby' (Williams 2005).

Pro-abortion activists tend to avoid talk of 'babies' and adopt the more technical terms when they are referring to unborn organisms. In their discourse they downplay the notion of the unborn entity as 'fully human' or 'fully a person', emphasising instead its undeveloped, imperfect body, its lack of full humanness. From this perspective, the unborn entity is only an unformed potentiality – a 'pregnancy' – rather than a fully realised individual. Given this ontological status of the unborn entity as 'not yet a person', for abortion advocates the pregnant woman's needs and desires are privileged over those of the 'pregnancy'. Using the term 'pregnancy' also positions the unborn as distinctly part of the maternal body, a physiological condition of her body, rather than configuring it as a separate entity with its own identity.

Anti-abortion activists, for their part, eschew using the terms 'embryo', 'foetus' or 'pregnancy' in favour of the 'unborn baby' as part of their goal to humanise the unborn. These activists view the unborn as fully human, as 'already babies', because they believe life and personhood to begin at the moment of conception. In anti-abortion discourses, therefore, terminating a 'pregnancy' is, quite simply, infanticide – or in more emotive terminology, 'the murder of a baby' (Ginsburg 1990, Hopkins *et al.* 2005, Taylor 2008). Indeed some commentators have adopted the term 'postnatal foetus' to describe all humans post-birth to demonstrate their antipathy towards abortion and their belief that the unborn are as equally as worthy of full moral personhood as any already born person (Hartouni 1991: 36). The use of the word 'adoption' to describe the donation of surplus embryos produced through IVF to other infertile couples similarly signals the status of embryos as already children (see more on this in Chapter 4).

So too, some medical facilities offer what they call 'foetal intensive care units' to monitor the health of foetuses that have been diagnosed with a medical condition. This title ignores the fact that the foetus is still part of the pregnant woman's body and must be monitored via her body (Casper 1994b). It is also interesting to note that on the 'pregnancy timeline' published on 'The Visible Embryo' website designed to demonstrate the weekly stages of development of embryos and foetuses, the

DOI: 10.1057/9781137310729

second trimester of pregnancy (gestational weeks 15 to 28) is entitled 'The fetal period', while the third trimester (weeks 29 to 40) is named 'The infant period', even though by definition this trimester is a term of pregnancy, not of the neonatal lifespan (*The Visible Embryo Pregnancy Timeline* 2011). All of these discursive and material practices contribute to the positioning of the foetus as already a child with its own autonomy and individuality.

Different countries and even regions within a country have different ways of distinguishing legally between a 'foetus' and an 'infant', based on such attributes as gestational age, size and whether the organism in question is considered 'viable' outside the uterus, with or without technological support. The concept of viability is integral to definitions between a 'foetus' and an 'infant' (Christoffersen-Deb 2012). The conceptual boundaries between these entities have become challenged and rendered unstable by medical technologies which are now able to facilitate the survival of premature infants at gestational ages at which they once invariably died, pushing back the medical and legal definition of viable life to earlier stages of development. The end of 24 weeks of gestation is often held to be the point at which viability may be established with neonatal care. Only a tiny proportion of infants born before this gestational age survive for long after birth. These 'previable' infants are often given 'comfort care' without being subjected to the heroic care measures to keep them alive that premature infants of greater gestational age are accorded (Chervenak and McCullough 2011). Once the 24-week mark has been passed – and in some hospitals, this distinction takes place literally at the stroke of midnight (Christoffersen-Deb 2012) – far more efforts are expended in maintaining the life of the premature newborn. After this point, however, determining viability is still subject to many observations, procedures and clinical decisions, often extending over weeks, that then structure what kind of care the preterm infant is offered and how its parents are counselled concerning its future prospects (Christoffersen-Deb 2012).

The definition of viability has also played an important role in many countries' legislation on elective abortion, including the USA and the UK. In both those countries, because 24 weeks' gestation is medically considered the point of viability, it is also the upper legal limit at which termination of pregnancy can take place, except under special circumstances (see Chapter 4 for further discussion of abortion policies across a range of countries). This places the foetus that is 24 weeks or older

DOI: 10.1057/9781137310729

in the ambiguous ontological position of being legally protected from abortion but still not yet born and not formally recognised as a 'person' in its own right (Christoffersen-Deb 2012). Given that the gestational age of the unborn in the late stages of pregnancy cannot necessarily be accurately defined within a margin of days or even weeks, even by ultrasound, this legislation itself rests upon assumptions of age. Thus, for example, obstetricians who conduct abortions may express disquiet about conducting terminations on foetuses that are estimated to be at least 20 weeks' gestational age, given this margin of error. Others may even be unwilling to terminate a pregnancy that is between 12 and 20 weeks' gestation because the foetus is closer to being viable, even though it definitely would not live if born at that point of development. As such, these doctors position the unborn as located on a timeline at which the 'natural' end point is birth and to intervene, therefore, is viewed as 'unnatural' and problematic (Beynon-Jones 2012).

Definitions of unborn entities and the language used to categorise them also play an important role in hospital and legal procedures around miscarriage and stillbirth. The methods of disposal of the remains of a pregnancy loss differ according to whether it has been determined to be an 'embryo', a 'foetus' or an 'infant'. Those entities that are designated 'infants' require a birth certificate and formal burial or cremation, while 'embryos' and 'foetuses' do not (Littlewood 1999, Williams 2005). As I discuss in more detail in Chapter 4, women who have experienced a miscarriage often find it difficult to accept that others often do not perceive their dead embryo or foetus as a 'real' person who deserves mourning and some kind of rite to mark its death. Definitions of probable viability also structure how pregnant women are classified and to which part of the hospital and under whose medical care they are sent. Pregnant women judged to be in premature labour with a foetus under 20 weeks' gestation are sent to the gynaecology ward, as it is assumed that the foetus will definitely not survive the birth, while those who are deemed to be giving birth to an infant (older than 20 weeks' gestation) are sent to the maternity ward (Christoffersen-Deb 2012, Littlewood 1999).

It is interesting to note that in past eras dead embryos and foetuses from miscarriages and abortions were routinely disposed of as garbage, and later as 'medical waste', with little concern for a respectful burial or disposal, regardless of their stage of development. It was not until civic municipalities began to be interested in sanitation concerns in the early years of the twentieth century that regulations for the disposal of these

DOI: 10.1057/9781137310729

entities began to be developed. Further, it was not until the late twentieth century, when the unborn began to be portrayed as individual persons as a result of the historical changes described earlier in this chapter, that this 'waste' began to be portrayed as 'human remains' requiring greater care in their disposal. No longer viewed as the responsibility solely of the state as part of sanitation programs, such groups as anti-abortion activists, people who had experienced pregnancy loss, members of the clergy and medical researchers involved in hESC science began to lay claim to decisions about how the dead unborn should be treated (Morgan 2002).

Further definitions around the unborn have been generated by the IVF, stem cell research and regenerative medicine industries. The cryogenically preserved embryo occupies a particularly ambiguous ontological state. It is not immediately clear whether this type of embryo is alive or dead, particularly if from a clinical perspective its development is viewed as 'arrested' via cryopreservation. Such embryos retain the potential to develop further if implanted in a uterus and thus their potential is suspended rather than being extinguished (Ellison and Karpin 2011). Like the donor cadaver, this type of embryo therefore occupies a marginal state of aliveness, not quite living and not quite dead, not quite human, not quite non-human (Ellison and Karpin 2011, Karpin 2006, Waldby and Squier 2003). This liminality is part of what makes decisions about how best to use or dispose of these bodies so contentious and difficult (see further discussion of this in Chapter 4).

The *ex vivo* embryo created solely for research purposes raises the spectre of a potential human life being brought into being just to serve instrumental purposes such as research or the provision of therapeutic material. This type of embryo may be contrasted with the spare IVF embryo, which is produced in the first instance for the purpose of assisting infertile couples to have children (Baylis 2000). The latter type of embryo has the potential for humanness, while the former is never able to realise its humanness because it is never to be implanted into a woman's body. The research-only embryo, therefore, is deprived of the capacity for humanness from the very start of its genesis. As this suggests, any embryonic body has the potential for personhood only via its connection to the female body. The *ex vivo* embryo deemed to be surplus or poor quality does not possess this potentiality. Aborted embryos have already lost their human potential once they have been taken out of the uterus. Thus not all embryos have equivalent moral status or value: they are contingent on female decision-making and

DOI: 10.1057/9781137310729

the embodied progress of pregnancy (Ellison and Karpin 2011, Karpin 2006, Rubin 2008, Waldby 2002a, 2002b).

So too, the ontology of the stem cells themselves which are derived from these bodies is ambiguous. Through their entry into the world of stem cell science, embryos are transformed from reproductive material with the potentiality for humanness into a radically different type of material with different potentiality: a stem cell line designed for research or therapeutic purposes with a spectrum of capabilities that can be extended into the future because it can regenerate apparently endlessly (Waldby and Cooper 2010, Waldby and Squier 2003). These cell lines therefore represent elements extracted from human bodies which may go on to have a future of their own, taking on a separate existence, just as donated organs do. They are not only alive; they are 'almost frighteningly' so because of this potential for immortality and their pluripotency (Waldby and Squier 2003: 35). Are the embryos created or used for the purposes of hESC science, and the biological material derived from them, commodities or persons? An entity can be a human or an object but not both under common law. Human tissue such as stem cells thus occupy an ambiguous space between object, and therefore potential commodity, and person, as they are neither fully one nor the other (Dickenson 2007).

For advocates of stem cell research, which is still very much in its experimental phase, the embryo is viewed as a precious resource. It is positioned as vital for developing stem cell research and therapeutic practice and thus offering the possibility for enhancing human lives and ameliorating suffering which otherwise would be wasted if disposed of. These advocates, in representing the embryo as a type of raw biological vitality, see the life of the embryo cells as being extended by being used to develop stem cell lines to treat others' illnesses: its vitality is diverted rather than extinguished (Rubin 2008, Waldby 2002b, Williams *et al.* 2008). Indeed one way in which scientific workers in hESC and regenerative medicine research may think of the embryos they use in their work is to see these entities' genetic potential as continuing on. According to this logic of genetic potential, although the embryo may be destroyed as part of the process of obtaining its useful cells, in some ways 'you have your genes and your stem cells growing somewhere forever' in other people's bodies, as one stem cell scientist put it (Ehrich *et al.* 2012: 119).

People with conditions that may be potentially amerliorated by stem cell therapy similarly position embryonic entities as highly valuable, and

DOI: 10.1057/9781137310729

indeed, as life-changing or life-saving in some cases (Ganchoff 2004, Langstrup and Sommerlund 2008). Here the embryo from which this material originated as well as the maternal body which produced it are both erased from view. The notion of the embryo as potential person is completely replaced by its positioning as therapeutic agent. The embryo is transformed into stem cell treatment, with the focus entirely upon other bodies – patients – and how they may be miraculously helped by these treatments. When such patients are themselves children, the discourse of potentiality and its accompanying meaning of value is transferred from the embryo to these children (Langstrup and Sommerlund 2008).

There is, therefore, a significant contradiction between two of the contemporary dominant definitions of the embryo: that which represents them as already fully persons and as already infants, and that which positions them as dehumanised therapeutic or research material. It is the conflict inherent in these definitions that is at the heart of the emotions and controversy aroused by these newer medico–scientific ways of treating and thinking about the unborn.

Concluding comments

This chapter has addressed the contingencies of unborn assemblages, emphasising that the ways in which we configure, think about and treat them are subject to change over time and across cultures and social worlds. I argued earlier that the advent of visualising technologies that were able to document the appearance of the unborn body – in particular foetal photography and obstetric ultrasound – was integral to configuring new ways of portraying this assemblage. The previously mysterious pregnant uterus was opened up to public gaze, and the embryo and foetus became public figures, entering new social worlds. The next chapter examines in further detail these technologies and others that serve to produce and reproduce images of the unborn and their implications for the meanings and practices that contribute to the contemporary unborn assemblage.

DOI: 10.1057/9781137310729

2
Imaging the Unborn

Abstract: *In the space of merely half a century, the unborn have emerged into the public spotlight from their previous dark and enigmatic domain. This chapter details the various ways in which visualising devices such as obstetric ultrasound technology, powerful microscopes, foetal photography and computer-generated imagery, as well as news and social media outlets, commercial commodities and museum displays have been used to depict and represent embryos and foetuses in both medical and popular culture. It addresses the commodification of the unborn image and examines how imagery has been used for political purposes, particularly in relation to anti-abortion activism.*

Key words: obstetric ultrasound; foetal photography; visualisations; media; embryos; foetuses; abortion politics

Lupton, Deborah. *The Social Worlds of the Unborn.*
Basingstoke: Palgrave Macmillan, 2013.
DOI: 10.1057/9781137310729.

Obstetric ultrasound: from medical diagnostic tool to 'baby pictures'

Prenatal care has undergone a major orientation towards the use of visualising technologies since the middle of the past century. In the 1960s, apart from x-rays, which were rarely used because of the harmful effects of radiation, no technologies were available to image, scan, test or screen unborn bodies. Obstetric ultrasound was first developed in the 1950s but was not introduced as a routine diagnostic technique until the late 1970s (Woo 2002). Since then, however, the obstetric ultrasound has become a rite of passage for the vast majority of pregnant women in developed and many developing countries. In these countries ultrasound scans are routinely offered to all pregnant women at around the 19–20 weeks' gestation mark to check on foetal morphology and development and to investigate evidence of any malformation or disability. Many women also have an earlier 'dating scan' towards the end of the first trimester of pregnancy to check on foetal wellbeing, growth and gestational age. Indeed in some countries, such as the UK, this 'dating scan' is recommended as routine for all pregnant women (Roberts and Thilaganathan 2007).

Developments in sonographic technologies have allowed for the capturing of embryonic development at its earliest stages. If it is deemed necessary, ultrasounds of embryos may now be undertaken using a transvaginal transducer (a tube-shaped probe inserted into the vagina). This is the preferred method until the eighth week of gestation, because it produces better images of the embryo than the more commonly employed abdominal ultrasound. Those women who have suffered recurring miscarriages, are experiencing vaginal bleeding or have a suspected multiple pregnancy because of medical treatment involving ovarian stimulation may undergo an ultrasound even earlier in the pregnancy (from the fifth week of gestation, when the conceptus may first be apparent on an ultrasound scan using a transducer) to check on the viability of the embryo, as may those women who have become pregnant via IVF (*Obstetric Ultrasound: A Comprehensive Guide* 2013).

Some countries offer many ultrasound scans as part of prenatal care. Germany, for example, has the highest rate of sonographic scanning in the world as part of its prescribed prenatal program. Pregnant Germans routinely undergo an ultrasound at each prenatal check-up, resulting in a total of 11 or 12 scans per woman by the time the infant is born (Erikson 2007, 2012). In some developing countries the unregulated

DOI: 10.1057/9781137310729

commodification of ultrasound technologies as part of private enter-prises allows pregnant women to undergo as many ultrasounds as they can afford. For example in Vietnam it is not uncommon for pregnant women to undergo as many as ten or more ultrasound scans during their pregnancies, even though the official Vietnamese Ministry of Health policy recommends scanning only for those pregnancies in which com-plications have been diagnosed (Gammeltoft and Nguyễn 2007).

Ultrasound images have been a major contributor in the inexorable trend towards the individualising and infantilising of the unborn. These images and the ways that they are discussed and interpreted by techni-cians and prospective parents construct a blurring of boundaries between the concepts of 'foetus' and 'infant'. Before the advent of routine obstetric ultrasound, the foetus did not become viewed as an 'infant' until birth. Ultrasound images now serve to render the foetus as an infant pre-birth and bring it into being as a body/self with which observers have a particular social and therefore ethical relationship. These images have therefore resulted in the 'social birth' of the new human to shift from the moment of physical separation from the maternal body at birth to earlier phases of unborn development, so that the bestowing of such social attributes as gender, personality and name often takes place before physical birth (Featherstone 2008, Han 2008, Mills 2008, Mitchell 2001, Morgan 1996, Taylor 2008).

Many services offering the relatively new technology of 3/4D 'social' ultrasound for imaging foetuses have emerged since the turn of this cen-tury. As the providers of these services are careful to warn their clients, these services are not oriented towards diagnostic imaging to identify any health or development problems in the unborn. They are solely for the purpose of allowing pregnant women and their partners and other fam-ily members to view images of the unborn as a social event. The services advertise that clients can take away such products as glossy photographs and DVDs of the images taken, set to music if desired, to include with their baby memorabilia (Kroløkke 2010, Roberts 2012, Taylor 2008). Many refer to the 'special bonding opportunity' their technologies offer. They typically and frequently refer to the unborn entity as 'your baby': one service has even called itself 'The Ultrasound Nursery'. Several providers advertise that they can bring their technology to the clients' homes and carry out the scan there. Some pregnant women have even begun holding 'ultrasound parties' employing these services, in which family members and friends are invited to their home to view the 3/4D

DOI: 10.1057/9781137310729

images of their foetus as a type of baby shower. This party may include a 'gender reveal' moment, where the sonographer identifies the sex of the foetus for the first time to parents and their guests, thus incorporating an element of surprise and additional entertainment (Bindley 2013).

Health professionals undertaking sonograms tend to adopt the discourse of infanthood when they are talking about the images with patients, thus contributing to this configuration. They often move between the technical terminology of the 'foetus', used mainly in written communication and medical notes, and that of the 'baby', used in conversation with each other and with the pregnant women with whom they are dealing (Mitchell and Georges 1997, Palmer 2009a, 2009b, Weir 1998). Sonographers, indeed, routinely make comments about the foetal images they view which ascribe to them personality, intentionality, human potentiality and social relationships and thus represent them as fully fledged 'babies'. They make reference to parts of the foetal body or face looking like features of the parents – their nose, eyes or long legs, for example – in an attempt to personalise the foetus, give it an identity and render it as part of the family. The foetus may be described as being 'shy', 'handsome' or 'pretty', 'not wanting his picture taken', 'athletic', 'smart', 'just like his Dad', 'very good' or 'cooperative'. Sonographers may speak to it directly as if it can hear and understand them, telling it to 'Smile for the camera' or 'Say hello to Mama' (Gammeltoft 2007b, Krøløkke 2010, Mitchell and Georges 1997, Palmer 2009a, 2009b, Roberts 2012).

The introduction of 3/4D ultrasound images have further contributed to this positioning of the foetus as already an infant, largely because they are able to show greater detail of the foetus's face, expressions and therefore supposed emotional state, interpreted as 'smiling', 'fear' or 'frowning' by onlookers. The focus in particular is on observing facial characteristics as part of bestowing an identity upon the foetus. In these services' information section it is frequently noted that they make an effort to capture the 'baby's face' in particular, but warn that '[y]our baby may hide behind hands and feet, cord, placenta or all of those mentioned' (*Mobile Ultrasounds* 2012). Juxtapositions of 3D images with photographs of the same newborn infant's face on elective ultrasound services' websites to show how 'true to life' the sonographic image is both serve to emphasise the 'realism' of the images and to highlight the concept of the foetus as baby, allegedly looking very similar *in utero* to the newborn self.

One Sydney-based non-diagnostic 3/4D ultrasound clinic put the following in their website blurb: 'Imagine seeing your baby smile, yawn or

DOI: 10.1057/9781137310729

stretch, whose nose does bubba have? Whose lips? With 3D ultrasound you can see all of this with great clarity.' It is further noted that the clinic prefers to take images at 24 weeks plus of gestation 'when bubba has a little body weight' – undertaken earlier, 'bubba can be a little skeletal and not give a true indication of what they really look like' (*Early View Ultrasound Centre* 2013). The use of the word 'bubba' to refer to the foetus and the emphasis on avoiding viewing the foetus when it looks too 'skeletal' (and therefore not 'baby-like') underpin the elision between foetus and infant which commonly takes place in such contexts. The tangible commodities that are produced – the ultrasound print-out or glossy photographs, the DVD showing the moving images – become objects that may come to stand in for the unborn entity itself, something that can be viewed and touched by others (Han 2008).

When they attend a 'social' 3/4D ultrasound session, expectant couples and their guests are offered a chance both to 'see' and to 'consume' the images of the foetus *qua* baby, positioned as eager tourists on a tour of the inside of the womb to view its marvellous contents (Kroløkke 2010, 2011). Sonographers who work for these services therefore move from the status of being allied health professionals in hospitals and obstetric clinics to acting as prenatal baby photographers. Indeed sonographers may represent their work as 'putting on a show' or 'giving a show' for the pregnant woman and other family members who may be present (Han 2008).

Ultrasound images, whether they are 2D, 3D or 4D, have had a profound effect on the ways in which pregnant women think about the body growing within them (see a detailed account of this in Chapter 3) and how they are treated in the medical setting. As noted in Chapter 1, the visual has overcome the haptic in contemporary understandings of unborn entities. The pregnant woman's embodied sensations relating to the unborn have lost currency, overwhelmed by the apparent precision offered by medico–scientific instruments. Ultrasound technicians check the embryo or foetus by looking at an externalised image on a monitor rather than touching the pregnant woman's body (Kukla 2005). Rather than medical professionals relying on the pregnant woman's account of what she feels in relation to the entity developing within her, her authority and knowledge are often superseded by the images produced by ultrasound. The dating scan undertaken in the first trimester of pregnancy is now often used to estimate the gestational age of the unborn, privileged over the pregnant woman's own knowledge of her menstrual cycle and

DOI: 10.1057/9781137310729

when she estimates that conception occurred. While doctors and mid-wives still measure the 'fundal height' of the pregnant woman's abdomen externally and may touch her abdomen during prenatal examinations to estimate the size of the foetus and its position, the images produced via ultrasound are considered far more accurate (Kukla 2005).

Photojournalism and computer visualisations

The technologies of medical photography and computer imaging have played an important role in bringing highly detailed and aestheticised portrayals of the unborn to popular consciousness. Before the advent of magnifying and ultrasound technologies able to show the embryonic body in detail, even women who had experienced early pregnancy loss would have had difficulty viewing the embryo they had lost because of its tiny size. However the moment of fertilisation of the human ovum by a sperm and the subsequent states of the development of the embryo as cells divide and multiply have now been depicted in great detail visually using powerful microscopic-facilitated photography, and more recently, computer imaging.

Lennart Nilsson's work capturing images of fertilisation and unborn development has been particularly influential from the mid-1960s onwards. Nilsson's most famous photographs of the unborn appeared first on the cover of the American magazine *Life* in 1965 and in a colour photographic essay entitled 'The Drama of Life Before Birth' published within, and appeared in various European magazines simultaneously. His images have also been published in glossy coffee table books, including perhaps his best-known work *A Child Is Born*, also published in 1965. These images, which were followed by several other books, a 1990 issue of *Life* and television documentaries, have been particularly influential in reconfiguring the visual portrayal of the unborn. They have attained an iconic status in portraying embryos and foetuses in full (albeit artificial) Technicolor glory.

To achieve his images Nilsson initially used macro-lenses light micro-scopy and then the scanning electron microscope, which were able to magnify sperm, ova and the early embryo many times to provide great detail. He incorporated 3D ultrasound into his technique when it was developed. The early magnified images, originally in black-and-white, were given full colour using techniques available at the time. Nilsson's

DOI: 10.1057/9781137310729

later images were colourised using digital imaging techniques. Most of the embryos and foetuses captured in Nilsson's photographs were dead specimens, although he did occasionally use a camera attached to an endoscope with a special wide-angled lens inserted in a pregnant woman's body to take images of living foetuses *in utero* (Nilsson's official website providing details of his techniques and displaying some of his images may be viewed here: *Lennart Nilsson Photography* 2013).

Another well-known professional producer of unborn images is the American Alexander Tsiaras, who runs a medical imaging company called Anatomical Travelogue (see here for his company's website: *TheVisualMD.com* 2013). In 2002 Tsiaras published an anatomical atlas of unborn development entitled *From Conception to Birth: A Life Unfolds*. The 'visualisations' of the unborn, as Tsiaras refers to his work, are created using computerised tomography scans and computer-enhanced magnetic resonance imaging. Tsiaras has since produced videos including these images such as 'From a Cell to a Baby' (2012) and 'Conception to Birth' (2011). 'Conception to Birth' was featured in a TED talk Tsiaras gave on his work (2011) in which he showed examples and repeatedly referred to the 'miraculous', 'mysterious' and 'magical' process that is the development of the unborn organism. Indeed he went so far as to comment that 'it would be hard not to attribute divinity to it'.

Both Nilsson's and Tsiaras's images represent unborn bodies as sublime, glowing like jewels in bright colours, figures of beauty and wonder. They are back-lit and artificially colourised to appear ethereal and appealing, set mostly against black backgrounds to set off their vividness (Morgan 2006c). The skin of the foetal bodies shown is the pink colour of newborn infants. The blackness in which the embryos and foetuses float and against which they are contrasted appears as the infinity of the universe, the unborn bodies and this space together representing the awesome power and mystery of life (an image which received particular resonance in Stanley Kubrick's film '2001: A Space Odyssey', which shows a giant late-term foetus floating serenely in outer space). The television documentaries made using Nilsson's work bore titles which also referred to the miraculous nature of conception and development of the unborn body to maturity, as in 'The Miracle of Life', 'The Odyssey of Life' and 'Life's Greatest Miracle' (Stormer 2008). Conception and development are represented in the works produced by Nilsson and Tsiaras as hazardous journeys in which danger lurks at every turn and many things can go wrong, halting progression (Stormer 2008: 664). Such images, in

DOI: 10.1057/9781137310729

concert with those produced by ultrasound technologies, contribute to the 'biotourism' approach to the foetal body, in which awe and wonder are inspired by the glorious vistas of inside the uterus, a mysterious and foreign space to which viewers are now able to have visual access (Krløkke 2010, 2011).

There is little if any indication of the maternal body in these images. For example, in Tsiaras's brief film 'From a Cell to a Baby' (2012), the unborn organism is shown rapidly developing through all the gestational stages from fertilisation onwards, isolated and alone against a black background. The film ends with the computer-generated imagery transforming into conventional video footage of an actual infant, sitting up and looking around (and therefore clearly several months older than newborn). Such representations not only suggest that unborn development is self-propelled and self-possessed, without the assistance of such vital organs as the umbilical cord and placenta. They also completely erase the maternal body from the birth process and present a frictionless trajectory in which the microscopic multi-celled organism becomes a months-old infant in seconds. The nine-month long work of the maternal body in conceiving, gestating and giving birth to the infant, and the interembodied nature of this work, completely disappears. It is as if the maternal body never existed and does not need to exist to produce the infant.

When the maternal body *is* acknowledged, it tends to be represented as merely as an 'environment' for the development of the precious unborn entity. It is telling that at one point towards the end of his TED talk Tsiaras (2011) displays one of his anatomical images of a pregnant woman's body, noting that she is 'a walking immunological, cardiovascular system that basically is a modal system that can actually nurture this child'. Here the pregnant body is configured as merely a receptacle for the wondrous miracle of new life that is developing within.

The unborn as cultural artefacts

The news media have played a significant role in contributing to the configuration of images of unborn entities since the initial days of embryo experimentation for the purposes of IVF in the 1970s. Given the widespread use of IVF in contemporary times, it is easy to forget the tenor of concern that was generated in some forums following its

DOI: 10.1057/9781137310729

initial use (as discussed in Chapter 1). IVF technologies aroused both rapturous coverage of the hope that they offered to infertile couples and negative portrayals about the ethical issues of creating embryos in the laboratory. References to Frankenstein's monsters and dystopian visions drawing upon science fiction were routinely made in the news media when representing the *ex vivo* embryo in the 1980s, accompanying concern over the regulation of scientific experimentation (Mulkay 1996). In the 1990s apparent cases of the wrong *ex vivo* embryo being implanted by accident, unrelated couples' gametes being used to make embryos and of gametes or embryos being deliberately 'stolen' from IVF patients and given to other couples made news headlines (Robertson 1995). Just as prevalent, however, were wonderful visions in news reports of the babies that could be made via IVF techniques and the joy that previously infertile couples experienced when they were able to conceive using IVF embryos (Birenbaum-Carmeli *et al.* 2000, Mulkay 1994, 1996).

As I observed in Chapter 1, the figure of the *ex vivo* embryo has become increasingly naturalised and domesticated now that over three decades have passed since the birth of Louise Brown. However it should be noted that IVF procedures tend to be accepted with the caveat that they are used for 'appropriate' couples: those who are considered young enough and are heterosexual. Women older than 50 years who bear children following IVF treatment, for example, still attract much moral censure and are portrayed as freakish and unnatural in the news media (Shaw and Giles 2009), while single women and lesbian couples are also often vilified in the media and parliamentary discourse for attempting to bear children via IVF (Michelle 2006, Smith 2003).

In these first years of the twenty-first century it is the newer technologies such as those employed by hESC scientists that are now receiving a high level of media attention. The *ex vivo* embryo that is specifically created for research purposes retains the status of the 'unnatural'. Deeply held fears and anxieties about how biomedical science may manipulate and alter human embryos are again evident in news media coverage of hESC science. The positions of those who view embryos as persons who should not be viewed as commodities or things and destroyed for the sake of bioscientific advancement and those who see them simply as organic matter which hold out therapeutic hope for people with serious medical conditions are constantly contrasted with each other in news accounts. The news media have also presented stories in which fears concerning uncontrolled science manipulating human embryos unethically and

DOI: 10.1057/9781137310729

unnaturally, creating such entities as clones, 'designer babies' or 'saviour siblings' (children created specifically to donate organs or cells to siblings affected with serious medical conditions), have dominated. In these representations, embryos become highly contested political sites as well as potentially monstrous bodies which have become altered via scientific procedures into forms which depart dramatically from accepted human bodies. Yet they also bear the meanings of hope, scientific progress and compassion for those with diseases or disabilities (Harvey 2005, Kitzinger and Williams 2005, Petersen 2002, 2005, Prainsack *et al.* 2008, Williams *et al.* 2003).

The emergence of such social media platforms as Facebook, Flickr, Instagram, Twitter, Bundlr and YouTube facilitating the sharing of images has allowed the wide dissemination of imagery and information about the unborn in public forums. Indeed sharing of the first ultrasound photograph on social media sites has become a rite of pregnancy for many women. This trend was lampooned by a cartoon in the *New Yorker* magazine appearing in December 2012, which showed a pregnant woman undergoing an ultrasound in a medical setting. The foetal ultrasound image on the screen was accompanied by the options to share it on Facebook, Twitter or YouTube. Some prospective parents sharing ultrasound images on social media choose to include ultrasound images highlighting such features as their foetuses' genitalia ('it's a boy!') or add captions to the images seeking to humanise and personalise the images '('peek a boo!') (STFU Parents 2012). Expectant parents can also add features such as the 'Baby Gaga' widget to their Facebook accounts that provide weekly updates on their unborn's development to their friends. Some have even set up separate Facebook and Twitter accounts for their unborn, posting updates on their behalf and in the process representing them as fully cognisant persons making statements about how much they love their 'mummy' and 'daddy' and 'can't wait to get here' (that is, be born) (STFU Parents 2011).

When the pregnancy of Kate Middleton, the Duchess of Cambridge, was officially announced in early December 2012, both the traditional news and social media coverage of the announcement configured what quickly became the most famous unborn entity in the world. Within hours several spoof Twitter accounts were created for this new individual, dubbed in the news and social media the 'royal baby' or the 'royal foetus', which apparently was tweeting from 'inside the royal womb' and giving commentary on its experiences. Depictions of the 'royal foetus'

DOI: 10.1057/9781137310729

used ultrasound images showing it already wearing a crown *in utero*. Various comments were made on Twitter concerning the wealth and social standing that the 'royal foetus' already enjoyed (Lupton 2012b). All this while the first trimester of the royal pregnancy had not yet elapsed.

There are many videos available for viewing on various pregnancy websites and YouTube showing fertilisation and embryonic and foetal development in brightly coloured 3/4D imagery, often set to portentous music and with a voice-over making reference to the dramatic nature of human conception and development. Several websites also provide detailed images of unborn development, often in a week-to-week format. One example is the website entitled 'The Visible Embryo' (*The Visible Embryo* 2012), which has as its emblem a spiral constructed of unborn life forms shown developing by week from the very first visible embryo once fertilisation has occurred to the fully gestated foetus ready to be born. Like the unborn in Nilsson's and Tsiaras's images, these entities are vibrantly coloured, positioned against a dark background. There is no evidence of the maternal body in view. Users can click on any of these images to see a larger image which demonstrates the appearance of the unborn organism and provides very detailed medico–scientific information about its anatomy. The images used are computer imaging models developed from the American National Institutes of Health collection of sectioned slides and specimens of embryos in addition to images captured from 3D and 4D ultrasounds of living foetuses.

YouTube includes many examples of ultrasound videos that have been uploaded by users wanting to share their good news and joy about their pregnancy. There are also numerous memorial websites and YouTube 'tribute' videos to lost pregnancies or stillborn infants on the internet (see further discussion of this in Chapter 4). Some of these images include ultrasounds at a very early stage of development, around the six-to-eight week gestational age mark, where the unborn body appears not much more than a tiny blob attached to the wall of the uterus. Several of these videos have been viewed by tens or even hundreds of thousands of viewers. These platforms, therefore, provide the technology by which private or commercial images of the unborn from a very early stage of development may be conveyed to a very large audience.

Earlier in this chapter I remarked on the trend towards offering commercial 3/4D 'bonding' or 'social' ultrasounds as entertainment for family members and friends and the commodities associated with this. Products such as Nilsson's and Tsiaras's videos and books are also

DOI: 10.1057/9781137310729

items that commodify unborn imagery, seeking to attract sales through their highly aestheticised representations of the 'wonder' of unborn development. An increasing number of other items are now available for purchase that use images of embryos and foetuses for decorative purposes. For example, consumers may buy from 'The Visible Embryo' website various products for personal use, such as t-shirts (including both maternity and regular-sized for women), caps, bags, hoodie jackets, sweatshirts, aprons, baby bibs, mugs, posters and computer mouse-pads, all emblazoned with the spiral image of showing the developmental cycle of the unborn. The use of the unborn as commodities and popular cultural icons ranges from jewellery, baby shower cakes in the shape of a foetus, t-shirts, Halloween outfits featuring foetal heads and limbs bursting bloodily out from a woman's body, advertising for both pro-abortion and anti-abortion organisations and groups, crocheted or knitted dolls, confectionary, soap and various artworks (see my Pinterest board for examples: *The Sociology of the Unborn* 2013).

An embryo has even featured as a character in an animated television series. The character of 'Embryo Princess' appears in two episodes of the Cartoon Network series *Adventure Time*. Although this character is named as an embryo, her appearance is that of a chubby, cute, pink and fully formed baby, with the exception of prominent red veins. She has a telekinetic power surrounding her like a big pink bubble that resembles an amniotic sac. She is an aborted embryo: the designer who created her originally wanted to name her 'Abortion Princess' but was forced to change the name because of resultant controversy (*Embryo Princess* 2012).

Images from ultrasounds have also become commodified and aestheticised in various ways. Ultrasound art magnifies and colours the ultrasound photograph and positions it on a large canvas to hang on a wall in one's home. People engaged in the craft of scrapbooking have used ultrasound photographs of their unborn in decorative displays. Special frames may be purchased for these images with sentimental messages emblazoned upon them ('Miracle in the making', 'Love at first sight'). Maternity t-shirts feature ultrasound images stretched across the protruding belly of the wearer. A new genre of professional maternity photography produces pictures of pregnant women holding ultrasound images of her unborn in front of her baby bump, or Photoshops the ultrasound directly onto the pregnant abdomen. Cookies and cupcakes have even been made with icing or toppers fashioned to look like

DOI: 10.1057/9781137310729

ultrasound images. Baby-shower invitations and thank-you cards, pregnancy announcement cards, fridge magnets, jewellery, lamp-shades and Christmas ornaments using ultrasound photographs are all available to purchase. (Examples of all these can be viewed on my Pinterest board *The Ultrasound as Cultural Artefact* 2013.)

These types of public representations of the unborn tend to be highly sanitised. Such images portray them as living and cute, ethereal and exquisite, a similar representation to the images of beatific sleeping infants that are so common in popular culture (Lupton 2013b, Morgan 2006c). Paradoxically, however, they rely upon previous dissection work performed by scientists that involved the use of dead embryos and foetuses. Without this 'gritty', 'grisly' and 'bloody' work, there would be no understanding of the appearance of the unborn body apart from that offered by the less detailed x-ray or ultrasound image (Morgan 2006c).

More realistic and less sentimentalised dead unborn bodies may be viewed in exhibitions such as 'BodyWorlds' (*BodyWorlds* 2013) where human body parts and cadavers with their skin removed are displayed in full anatomical detail. The various 'BodyWorlds' exhibitions, which have toured the world since the early 2000s, have brought the unborn cadaver from the medical school out into public view. Some of these exhibitions have included a section with a display of embryos and foetuses at various stages of gestational development, including those with morphological abnormalities. They have also displayed the body of a pregnant woman with her skin removed and her abdomen cut open to display the eight-month-old foetus within. All of these are medical specimens or cadavers which have been donated to science and treated with the process of plastination (in which the water and fat in a cadaver or body part are drained away and replaced with polymer plastic materials that preserve the specimens).

The display of these plastinated unborn cadavers has caused some consternation and controversy, with some commentators arguing that putting dead human bodies – and particularly those of unborn entities – on display was repellent. One English newspaper even used the headline 'Flayed babies' bodies included in new Body World [sic] exhibition' when the exhibition was displayed in the UK (Adams 2008). It is perhaps because of the anticipated controversy provoked by displaying preserved embryonic and foetal specimens that the section featuring them was separated from the rest of the exhibition, screened behind a curtain. The suggestion of such positioning is that these specimens

DOI: 10.1057/9781137310729

are different from cadavers of older humans, presumably because they are of 'dead babies' and therefore should be accorded some degree of extra care or privacy because of sensitivities and sentimentality around infants. Obvious images of dead unborn bodies may also be found in websites such as 'The Multi Dimensional Human Embryo' (*The Multi Dimensional Human Embryo* 2009). This website displays each stage of development in great detail, employing 3D imagery using MRI technology as well as images of embryonic internal organs, obviously using dead embryos to do so.

Many people find viewing obviously dead unborn specimens an uncomfortable experience, because they do not display the 'cute', highly aestheticised attributes of most other visual representations of embryos and foetuses (Morgan 2006c, 2009). As used as we are to the artificially manipulated visualisations presented in popular culture, the 'reality' of the unborn specimen is confronting of the sentimentality that surrounds the unborn. Photographic images of dead embryos or foetuses or actual specimens that are not re-touched, artificially coloured or computer-generated are far less 'cute' and infant-like. For this reason, photographers and computer artists attempting to configure the unborn as aesthetic objects have manipulated their images to render them more beautiful and infantile. Indeed Tsiaras's unborn visualisations were market-tested with consumers to ensure their maximum appeal to the buying public before his book was published (Morgan 2006c).

The politics of unborn imagery

As many writers have noted, particularly those adopting a critical feminist perspective, the visual conventions of contemporary images of the unborn contribute to an ontological separation of the unborn organism and its needs from that of the woman who is gestating it (see, for example, Adams 1993, Duden 1993, Hartouni 1991, Maher 2002, Mitchell 2001, Morgan 2009, Palmer 2009a, Petchesky 1987, Stabile 1992, Taylor 2008). As I have shown earlier in this chapter, the dominant convention of representation of the unborn assemblage is that which portrays the unborn body floating in a dark chamber that is not recognisable as a woman's uterus, frequently with little or no evidence of the umbilical cord or placenta. This visual approach is evident from depictions of the moment of conception onwards throughout the development span of

DOI: 10.1057/9781137310729

the unborn organism. The embryo is shown as miraculously developing with all its organs and appendages and other physiological systems in the complex process of producing a human life. It is common for unborn entities to be represented as achieving this process independently, with little or no help from the maternal bodies in which they reside, which are often routinely cropped out of the picture (and out of the voice-over narrative of films).

In some ways it is perhaps not surprising that the visual separation of the unborn from the maternal assemblage has become a dominant cultural archetype. Despite the presentation of the images as being of living creatures, most of these embryos and foetuses portrayed in commercial photography and online websites are actually dead bodies taken from the cadavers of pregnant women or from tubal pregnancies and preserved as part of the collections of medical schools or science institutes. 'The Visible Embryo' website's images, for example, came from the American Carnegie Institution of Science's extensive collection of cross-section slides of embryo specimens rendered into 3D form by sophisticated computer programs. Nilsson's and Tsiaras's photographs were also made using mostly dead specimens that were posed and technically altered to look particularly endearing, attractive and vital (Morgan 2006c). The maternal body was absent because these unborn bodies were not actually still positioned *in utero* by the time they were used to make the images. However even when a living foetus was photographed by Nilsson using an endoscopic camera inserted inside a pregnant woman's body, visual evidence of her body was removed from the picture in the resultant publications, achieving the same visual effect of autonomy of the unborn entity. This convention of unborn representation and our expectations that it is met has resulted in umbilical cords and placentas being routinely cropped from ultrasound images, so that prospective parents can display the image of their unborn without the 'messiness' of such reminders of the maternal body (Nash 2007, Palmer 2009a).

Such images have been used repeatedly in anti-abortion materials to promote the notions that life begins at conception and that the earliest unborn organisms, the zygote and blastocyst, are simply the first stage in the development process leading to a fully realised, unique human being (Boucher 2004). The separation of the unborn entity from that of the body in which it is growing, the infantilisation of this entity, its portrayal as beautiful, vulnerable and precious and its very status as an entity of which 'baby photos' may be taken, have allowed anti-abortion

DOI: 10.1057/9781137310729

groups to use these images to promote their objections to pregnancy terminations. Indeed anti-abortion activists argue that women who are seeking a termination of pregnancy should be shown ultrasound images of their embryo/foetus, as this visual image will allegedly persuade them to view it as their 'baby' and they will then decide against a termination (Hopkins *et al.* 2005). It is already the case in some American states that women wishing to undergo a termination of pregnancy are by law first obliged to view ultrasound images of their unborn, as part of confronting them with the 'reality' of the body developing inside them and forcing them to acknowledge its state of development and possible humanness (Han 2008, Mills 2008). Some Christian groups in the USA who oppose abortion and offer support to women with unwanted pregnancies have even begun to set up their own ultrasound clinics for this purpose alone (Han 2008).

The images made available by 3/4D ultrasound technologies have been considered in a parliamentary committee in the UK in relation to abortion legislation and legal limits concerning at what gestational age foetuses can be legally aborted, based on supposed foetal conscious-ness and sentience and the level of development akin to infancy dem-onstrated in these images (Mills 2008, Palmer 2009b). In speaking to that committee, the British obstetrician and ultrasound pioneer Stuart Campbell used 3/4D images taken in his practice to contend that there should be a reconsideration of the gestational time limit for abortion in that country and that it should be reduced from 24 to 12 weeks' gestation (Palmer 2009b). Campbell was widely quoted in the British news media as making such statements as: 'The more I study foetuses the more I find it distressing to terminate babies who are so advanced in terms of human behaviour' (Palmer 2009b: 175). It is notable, however, that Campbell was more critical of aborting what he called 'normal babies' for 'social reasons' than, by implication, 'abnormal' foetuses (Palmer 2009b).

The first well-known film using ultrasound images to be made as a persuasive tool on behalf of anti-abortion groups was the American 'The Silent Scream', to which I referred in Chapter 1. This film purported to show, in real-time, the experiences of a 12-week foetus going through an elective abortion. The film was narrated by an ex-abortionist physician, who describes the foetus shown as becoming agitated and anguished as it allegedly 'senses danger' and attempts to escape from the aspira-tor used to suction it from the woman's uterus. The narrator claims at one stage that the foetus is emitting 'a silent scream' of fear and pain

(Hartouni 1991: 36–37). In a more recent film, 'Ultrasound: Window to the Womb' (1990), which was produced for an anti-abortion campaign, moving images of foetal bodies from ultrasounds are shown, with the narrator interpreting the body parts and movements for the audience to personalise the foetus and render it child-like and responsive to its audience. The narrator uses such expressions as 'he jumps', 'he just waved at you', 'now he's turning and looking at us' to describe the actions of a ten-week foetus (Boucher 2004). Such representations suggest that the foetus, even at this very early stage of development, has already acquired personality, intentionality, social relationships, free will and individuality independent of its mother.

Nilsson's and Tsiaras's images have also been frequently used by Christians and anti-abortionists to support their beliefs that unborn life is sacrosanct and should be protected at all costs (Morgan 2006c). This interpretation has been encouraged by the representation in their work of human conception and unborn development as miraculous and wondrously complex. Tsiaras' predilection for highlighting and magnifying the hands, feet and toes of embryos and early foetuses in his visualisations serves to emphasise their apparent infant appearance, thus supporting anti-abortion activists' claims that unborn entities are already 'babies' (Morgan 2006c). Several anti-abortion websites and printed materials offer gruesome photographs of aborted embryos and foetuses, many of which, according to some pro-abortion activists, are fake or use misrepresentations of embryonic/foetal age. These are often juxtaposed with photographs in the style of Nilsson's images, contrasting beautiful, serene and softly lit images of the embryo or foetus with harshly lit grimly realistic photographs of their dismembered, bloodied bodies allegedly resulting from abortion – referred to by anti-abortion activists as 'war pictures' (Ginsburg 1990: 104).

Other contemporary anti-abortion media materials use such social media platforms as YouTube to position the unborn as vulnerable humans requiring protection from the death that awaits them. Thus, for example, one of the multitude of such videos uploaded to YouTube, 'Sad Abortion Video' (Anonymous 2012), uses a combination of infantilising the embryo, rendering into a human capable of thought and speech, representing it as a newborn infant and small child by giving it a name and predicting its future appearance and interests. Words are used from the perspective of the unborn talking to its mother: for example, 'Hi Mommy! I'm your baby. You don't know me yet, I'm only a few weeks old ... My name is John and

DOI: 10.1057/9781137310729

I've got beautiful brown eyes and black hair'. Later in the video as the abortion is supposedly taking place, the foetus is purportedly saying, 'I'm really, really scared Mommy. Please tell me you love me... Please make them stop! It feels bad... Please, Mommy, please, please help me' and so on. This highly emotive language giving an unwarranted agency and consciousness to the unborn entity is accompanied by images from 2D and 3/4D ultrasounds (one supposedly showing a foetus crying), foetal figurines, newborn babies sweetly sleeping and laughing and, at the end of the video, the bloodied bodies of terminated foetuses. This video, by September 2012, had been viewed well over 1 million times and had attracted 18,000 comments, many of which were from people noting the poignancy of the video, their distressed feelings in response and condemning abortion.

Other anti-abortion products depicting the embryonic and foetal form are the dolls produced by the company 'God's Little Ones', whose motto reads 'Hold the beauty of life in the palm of your hand!' (*God's Little Ones* 2012). According to the website, these dolls 'prove that the unborn are human beings deserving love and respect'. Life-size embryonic and foetal models (described with the infantilising term 'micropreemies') from the earliest visible stages of development are produced by this company, which also supplies display cases advertised as a 'great tool for sidewalk counselling'. The intention of this company and its products to advocate for anti-abortion politics is evident from this emotive discourse, in conjunction with products such as 'pro-life pins' which show 'eight-weeks gestation micropreemie' dolls placed in the middle of plastic flowers (a 'great gift for a pro-life volunteer'), as well as infant angel figures cuddling embryos and eight-weeks-gestation foetuses placed next to gummy bear sweets on a pin to demonstrate their tiny size. Other life-size embryo and foetus dolls from six weeks gestation upwards show them wearing little hats, vests and nappies and with teddy bears, serving to represent them as equal to infants in their value and emotive appeal. In appearance the embryo and early foetus dolls are far more similar to infants than they are to real unborn entities, with the pink skin of infants rather than the livid red skin of embryos and early foetuses.

Concluding comments

Various themes have emerged from this chapter concerning the importance of visualising technologies, the news and social media and popular

DOI: 10.1057/9781137310729

and commodity culture in configuring the unborn. These include the integral role played by these technologies, practices, objects and forums in rendering the unborn assemblage as independent, separate from the maternal assemblage, evincing their own personhood and personality well before the time of birth; the ways in which unborn images produced by technologies such as ultrasound that were originally designed for medical use have expanded beyond the social world of the clinic into the public domain of commodity culture; the transformation of figures of the unborn into popular cultural artefacts and commodities; the dominant tendency to aestheticise and infantilise representations of the unborn in visual medical and popular culture; and the intensely political nature of unborn images, particularly as they have been taken up and used in anti-abortion activist materials. The next chapter goes on to review the ways in which pregnant women themselves conceptualise and experience the unborn entities growing within them.

DOI: 10.1057/9781137310729

3

The Unborn within the Self: Women's Experiences of Pregnancy

Abstract: *This chapter brings the maternal body back in to view by focusing on the embodied experiences of pregnancy. To do so, I draw first on feminist sociological and philosophical inquiries into the ontology of pregnant embodiment and then on empirical research that has sought to elicit women's experiences of pregnancy and their concepts of the unborn growing within them. The chapter discusses the ambiguities and ambivalences of pregnancy and how women may struggle with coming to terms with harbouring another body within their own. The role of imaging technologies in contributing to women's experiences of their unborn is reviewed, as are the complexities of the concept of the maternal–foetal 'bond'.*

Key words: sociology; feminist philosophy: experiences of pregnancy; foetuses; embodiment; Self/Other

Lupton, Deborah. *The Social Worlds of the Unborn.* Basingstoke: Palgrave Macmillan, 2013. DOI: 10.1057/9781137310729.

DOI: 10.1057/9781137310729

Ambiguities of pregnancy: the two-in-one body

Concepts of unborn subjectivity and embodiment are necessarily linked to notions of adult subjectivity and embodiment. Western notions of mind and body divide them and represent them as different entities, with the mind taking precedence, viewed as being able to exert control over the flesh (Conklin and Morgan 1996, Grosz 1994, Longhurst 2000a, Shildrick 1997). Contemporary western societies place an emphasis on individuality, self-containment and autonomy, with little recognition that social interactions and relationships contribute to the creation and maintenance of one's body. One's own body is primarily viewed as entirely separate from others' bodies. It is one's private property, to be regulated, disciplined and controlled by oneself. This concept differs radically from the relational model of personhood and embodiment common in non-western cultures, which depends on creating and maintaining social ties with other persons/bodies and which views bodies as communal rather than as individuated from each other (Conklin and Morgan 1996, Kaufman and Morgan 2005).

In western cultures the unborn body and the pregnant body are thus both anomalies according to accepted norms of 'proper' individuated and contained embodiment. They are ambiguous and unsettling in their blurring of boundaries between Self and Other. Where does the pregnant body/self begin and the unborn body/self end? Given its ambiguity of boundaries, the pregnant body is symbolically leaky, permeable and porous (Grosz 1994, Hird 2007, Kristeva 1982, Kukla 2005, Longhurst 2005, Maher 2002, Schmied and Lupton 2001). Unlike any other body, it contains within it another human body that eventually must be expelled to split the two-in-one body – the unborn–maternal assemblage – and render it two separate bodies. It challenges the notion of the liberal human subject as individuated from others, and of the 'proper' body as separate from other bodies, tightly contained, its borders rigorously policed and defended (Betterton 2002, Ruddick 2007). Indeed for the feminist philosopher Julia Kristeva (1982) the pregnant body is the epitome of the abject body – the body that inspires revulsion but also fascination – because of the blurred subjectivity of the mother/unborn body/self, its ambiguous status of personhood.

Thirty years ago the feminist philosopher Iris Marion Young (1984) wrote an influential analysis of the ontology of pregnant embodiment. In her essay Young begins by drawing attention to the ways in which

DOI: 10.1057/9781137310729

medical journals and pregnancy handbooks tend to represent pregnancy as a physical state of the developing foetus, an objective scientific process for medical study or a 'condition' of the pregnant woman. She observes that accounts of the phenomenology of pregnancy are lacking in this literature: that is, how it feels to be pregnant in relation to concepts of selfhood and embodiment. Young goes on to note the lack of distinction between Self and Other that pregnant women often experience. Drawing upon the writings of Kristeva on pregnancy, she contends that the pregnant subject challenges the concept of the unified subject because she 'is decentered, split, or doubled in several ways. She experiences her body as herself and not herself' (1984: 46) Young contends that the pregnant woman's sense of her body boundaries change as her body changes shape, which has implications for her subjectivity. As pregnancy progresses, she becomes more aware of her body and less sure of where her body/self ends and that of the unborn begins.

Liminality of body boundaries creates cultural imperatives to control and contain such ambiguity (Douglas 1969, Grosz 1994, Kristeva 1982). Popular portrayals of the pregnant body tend to be ambivalent and contradictory. Pregnant women are represented as engaging in the most worthwhile pursuit a woman can achieve: gestating a new human. As long as the pregnant woman conforms to societal norms and expectations of what a 'pregnant woman' should be (that is, not too old or too young, ideally married or at least in a stable heterosexual relationship, not already the mother of too many children), pregnancy is revered and honoured for this reason. To conform to the norms and ideals of pregnancy, pregnant women are expected to adopt the 'madonna' style of impending motherhood, behaving with decorum and heightened self-discipline, presenting a motherly, asexual self (Longhurst 2000b, 2005).

More negatively the pregnant body is commonly portrayed as ruled by wildly fluctuating hormones, out-of-control, experiencing strange food cravings or simply eating too much, liable to the uncontrolled expulsion of bodily fluids and even as hyper-sexual. The 2012 Hollywood film 'What to Expect When You're Expecting', for example, made much of the lack of control pregnant women have over their bodies and their propensity to be highly emotional, vomit in public places, have their 'waters' break unexpectedly or to lack bladder control. One of the pregnant female protagonists in the film at one point cries out in distress, 'I have no control over my body or my emotions'. Such portrayals show the supposedly inferior capacity that women – and particularly those who are experiencing such physical states

DOI: 10.1057/9781137310729

as pregnancy or childbirth – have in regulating and disciplining their bodies in the ways expected of the ideal rational and contained subject (Draper 2003, Longhurst 1997, 2005, Schmied and Lupton 2001).

The pregnant body is extremely powerful in its ability to produce another human. This power, exclusive as it is to women, has both positive and negative implications. As feminist writers from the 1970s onwards have noted, women's reproductive capacities may serve to position them as merely or primarily reproducers of human life: as vessels for procreation, as breeding bodies that are less capable of rational thought and action than other human subjects. They have contended that the association of the reproductive body with femininity, and assumptions that maternity is a 'proper' role for all women, may act to oppress women and deny them opportunities to go beyond this role (Ehrenreich and English 1974, Hird 2007, Longhurst 2005, Martin 1992, Oakley 1984, Tyler 2009). The power of the maternal body can also be experienced as engulfing and overwhelming on a psychoanalytic level, as Kristeva (1982) argues, contributing to negative feelings about the maternal subject and the dependency that one has on this subject. Kristeva claims that the first object of abjection for the child is the maternal body, from which the child seeks to individuate itself, creating a border that previously did not exist between Self and the maternal Other.

Yet women's procreative capacity may also bestow upon them a sense of superiority, meanings that often surround the archetype of the 'good mother'. Women may achieve intense feelings of empowerment, meaning and accomplishment in gestating, giving birth to and caring for their children. The size and bulk of their pregnant bodies may also bestow a sense of power and heightened awareness of sexuality and of others' approval that they are fulfilling their maternal function (Young 1984). Part of the procreative experience for women may be awareness of the positive dimensions of the interconnectedness and interembodied nature of pregnancy and nurturing for children, the lack of individuation and the pleasures of permeability of selfhood and embodiment (Hird 2007, Stephens 2005, Young 1984).

How pregnant women conceptualise the unborn

Since Young wrote her essay on the phenomenology of pregnancy, noting the lack of writing on this topic, several empirical studies have been

DOI: 10.1057/9781137310729

published that have investigated women's experiences of pregnancy. This research has highlighted the fluidities and ambivalences of pregnant embodiment identified by Young. It shows that distinctions between Self and Other may change at different points of time in the pregnancy. Women sometimes feel as if their unborn is part of them but at other times position the unborn as an Other to their Selves. They may move back and forth between these positions.

Pregnant women often have difficulties reconciling themselves to the notion that there is another body inside their own (Longhurst 2005, Schmied and Lupton 2001, Young 1984). This may particularly be the case in the early stages of pregnancy, when no 'baby bump' is obviously apparent and the woman does not yet feel any foetal movements (Nash 2013). At this stage, pregnant women become aware of heightened sensations, emotions and physical changes in their bodies, but many find it difficult to conceptualise or articulate the ontology of pregnant embodiment. For many women the first movements of the foetus ('quickening' as the traditional term puts it) is the point at which they feel a sense of their subjectivity splitting into two parts: their own and that of the foetus (Nash 2013, Root and Browner 2001, Schmied and Lupton 2001). These movements may feel strange because they are both within one's own body and instigated by another's body (Young 1984). Some women in the later stages of pregnancy have described the movements of the foetus as 'weird' because they are so different from the usual state of (non-pregnant) embodiment (Nash 2013).

In an interview study with Australian women experiencing their first pregnancy that I conducted with Virginia Schmied (Lupton 1999, Schmied and Lupton 2001), we found that the women struggled to articulate their experiences of gestating and carrying their unborn. They used such phrases as 'I don't know how to put it into words', 'It just blows my mind' and 'I can't explain it' when attempting to describe how they felt about the embodied state of pregnancy. When describing the unborn inside them, they made such comments as: 'I just can't think of the right words to describe what it feels like when it moves', 'You feel like you've got a constant little companion' but at the same time 'It's as though it's part of your body'. When the women discussed their experiences of late pregnancy, the unborn body was still described as 'part of your body', but also as 'developing its own personality. It's becoming, I suppose, less and less dependent on me or less and less a part of me and more an individual', as one woman put it (Schmied and Lupton 2001: 35). These

DOI: 10.1057/9781137310729

concepts were also evident in Nash's (2013) interviews with Australian women, in which the participants were roughly equally divided between those who viewed their foetuses as 'part of me' and those who thought of them as separate individuals.

Despite the tangible evidence of another growing and moving body within their own as pregnancy progresses, such as feeling the unborn body's movements and witnessing their abdomens becoming gradually larger, pregnant women often still struggle to come to terms with the idea that an infant – another person – will eventually emerge from their own bodies. As one of the woman in our study noted, '[the foetus] is something that's hard to see, something that's part of you. And then [after birth] it's going to be something out here and you sort of wonder what happened between the two.' Another woman was having difficulty conceptualising of her unborn as separate from herself even in late pregnancy: 'I still can't see it as a separate being to me. I've tried, but it hasn't worked' (Schmied and Lupton 2001: 35–6). Her words suggests that there is a cultural expectation that pregnant women should reach a point, particularly in late pregnancy when birth is imminent, when they are able to view their unborn as autonomous individuals.

Prevailing popular discourses on pregnancy often represent it as a wonderful experience for women, allowing them to revel in the miracle their body is producing in gestating another human and to experience the joys of 'becoming a mother' (Gentile 2011, Longhurst 2005). In such a context it may be very difficult for women to express negative feelings or even a degree of ambivalence about the unborn entity growing within them, at least to family members or friends. Yet interviews with pregnant women from such countries as Australia, the UK and the USA have found that for many women, pregnant embodiment is experienced as losing control over their bodies, with the unborn dictating their bodily dimensions and sensations. Some women find this status unsettling in its ambiguity. They may therefore conceptualise the unborn entity as an antagonist who is controlling their own bodies. Some women even find the experience of pregnancy like an invasion, viewing their bodies as being taken over by an 'alien thing' and expressing the idea that their body therefore 'no longer belongs' to them (Warren and Brewis 2004: 223). As one Australian woman in our study said, 'The most amazing thing for me was that something is growing inside me that I can't control at all. It's an invasion of my own body. In the beginning, I'm sure I used words such as "parasite".' Another participant in our study made a

DOI: 10.1057/9781137310729

similar reference to 'something using you as a host…. I just felt a real loss of identity and autonomy' (Schmied and Lupton 2001: 36).

While they may not talk about 'invasion' by 'parasites', other women have articulated the idea that the unborn is negatively affecting their health, wellbeing or physical appearance. Pregnant women in one American study described how their unborn made them feel tired, irritable, hungry or nauseated. The unborn entity was conceptualised as 'taking nutrition' from the mother's body to the detriment of the woman unless she was careful to eat enough (Root and Browner 2001). In another American study, women also expressed their awareness that the discourse of 'maternal sacrifice' meant that it was difficult for them to complain to others about the physical discomforts, pain or illness they experienced as a result of pregnancy. It is expected that because these symptoms are a result of becoming a mother, and that this is a par- ticularly 'special' and 'fulfilling' time for women, then pregnant women should not complain but rather focus on the wellbeing of their unborn and the 'prize' of having a newborn after having endured this suffering (Bessett 2010).

In British research, the interviewees talked about the changes in their bodies' size during pregnancy. Some women were proud of their large 'bump' as a symbol of their impending motherhood and felt more 'wom- anly' with their more rounded shape. However other women disliked looking 'fat' and the experience of developing larger breasts and rotund abdomens, and were looking forward to returning to their usual body size and shape following the birth (Earle 2003). Similar findings were evident in Australian research which found that women who enjoyed engaging in rigorous fitness routines to keep fit and slim found it dif- ficult to relinquish them during pregnancy. Many said that they disliked feeling 'fat' or being described by others as 'fat' when they were merely pregnant (Nash 2013).

Feelings of hostility, of viewing the unborn as an invader or a parasite, are often evident in the accounts of women who are experiencing an unwanted pregnancy, particularly if they have become pregnant from an act of rape. In these situations, because the unborn is a product of violence and is genetically part of the perpetuator of this violence, preg- nant women often conceptualise the entity within them as monstrous, less than human, even as a kind of cancerous growth, and certainly not as an 'innocent baby'. They may view their pregnant bodies and the unborn contained with them as 'disgusting'. If the rape was part of war, the notion that a captor's or invader's genetic material is gestating within them can be horrifying for such women. They may feel as if their own bodies have become battlegrounds. Here the unborn is conceptualised and experienced in unambiguous terms as a foreign and repellent Other and most definitely not as part of the Self (Lundquist 2008).

Many women, however, enjoy the sensation of harbouring the unborn body within their own and experience the loss of control over their bodies almost as a 'relief', a temporary reprieve from needing to constantly discipline their bodies or worry about being 'fat' (Nash 2013). These women tend to conceptualise their unborn more positively and find the generation of this Other body within their own as a miraculous and awe-inspiring process (Carter 2010, Longhurst 2000a, Root and Browner 2001, Warren and Brewis 2004, Young 1984). In the research study Schmied and I conducted, for example, several women talked about their unborn as 'constant companions' and made reference to enjoying the warm embodied sensations of 'cuddling up' to them (Schmied and Lupton 2001). While this type of discourse also positions the unborn as an Other body to one's Self, this is in a far more positive light compared to the concept of the antagonistic or parasitic unborn Other.

Concepts denoting an emotional separation from the unborn entity are also commonly expressed by women in childbirth, the time when the maternal body expels the Other's body and they are physically individuated from each other once the umbilical cord has been cut. In another phase of my research with Schmied (Lupton and Schmied 2013), we examined the sections from our interviews when the participants were discussing giving birth to their first infants, looking for their descriptions of the moment of birth and how they talked about their own bodies and those of the bodies they were expelling from their own. We used the term the 'body-being-born' to denote its ambiguous state of being both inside the woman's body and also in the process of coming out. This 'body-being-born' is, of course, another kind of unborn–maternal assemblage. We found that towards the end of labour some women described coming for the first time to an awareness of the body-being-born as a separate body from their own. However, this awareness was often partial. During this period the body-being-born may emerge and re-emerge from the labouring woman's body for some time as the head starts to push through. This period of slippage of the body-being-born from 'inside' to 'outside' and back again, over and over, is extremely ambiguous in terms of the containment of the labouring woman's body. It is also ambiguous for

DOI: 10.1057/9781137310729

definitions of this body as foetus or infant. According to technical medical definitions, when it is still inside the woman's body this body remains a foetus. Once it is fully outside, it is an infant. Yet technical terms do not fully encapsulate the labouring woman's physical and conceptual experiences that during this transitional phase, the body-being-born is liminal, neither fully one nor the other.

The process of emergence may also challenge labouring women with the foreign sensation of parts of another's body protruding from one's own (see also Young 1984). When the women in our study talked about the moment of birth, they often described the strangeness of feeling the body-being-born emerging from their own body. This was confronting for several women. One of our interviewees recounted how she was asked by the midwife whether she wanted to touch the infant's head as it emerged from the birth canal: 'I didn't feel like it – I suppose I was a bit scared of the unknown.' As another woman commented of the moments after birth when she was holding her infant in her arms and gazing at it, 'It was like looking at something, wondering "Where did this baby come from?" Despite what I'd gone through it was hard to associate that she was actually mine and she was out of my stomach.' As these comments suggest, for many women in labour or who have just given birth, the 'body-being-born' can be conceptualised as alien, strange, foreign to oneself and one's body. It can take some time for women who have just given birth to accustom themselves physically and mentally to the notion that another body has just emerged from their own, from which they are now individuated.

The role of imaging technologies in pregnant subjectivity

In her essay on pregnant embodiment, Young discussed the uniqueness of the pregnant woman's haptic experiences of the unborn. Describing the sensations of foetal movement from the perspective of a pregnant woman, she writes: 'Only I have access to these movements from their origin, as it were. For months, only I can witness this life inside me ... I have a privileged relation to this other life' (Young 1984: 48). Young comments that pregnant women may feel alienated from their experiences by their encounters with obstetric medicine because pregnancy is often treated as a 'disorder' that requires medical attention. She makes reference to 'medical instruments'

DOI: 10.1057/9781137310729

that 'objectify internal process' so that the woman's experiences are devalued (1984: 55) and argues that technologies directed at the foetus tend to ignore the pregnant woman's knowledge of her body. However Young is referring mainly to the medical technologies used in labour, such as foetal monitors. She was writing before ultrasound technology had begun to make a significant impact on women's experiences of pregnancy – in fact she makes only a passing reference to ultrasound in her analysis. Indeed it is striking that Young's phenomenological account of pregnancy echoes Duden's (1993) descriptions of the ways in which the unborn were experienced in early modern times (discussed in Chapter 1).

In the period of almost three decades that has elapsed since the initial publication of Young's essay, the role of ultrasound in defining the pregnant experience has intensified. Most women no longer have the privileged access to the unborn of which she writes. This access is now shared with the visual image produced by ultrasound, photography and computerised imaging technologies. It has been contended that these images and others like it in popular culture have become so pervasive that pregnant women have come to conceptualise their uteruses and the unborn using such images. They thus have developed a third-person relationship to their wombs, drawing on 'a single, canonical fetal figure' (Kukla 2005: 288) rather than conceptualising their own embryo or foetus as differentiated from this figure.

Feminist academic Alice Adams (1993) has written of the influence of Lennart Nilsson's book *A Child is Born* upon her own experiences of pregnancy. She notes that the images in the book of the embryonic and foetal bodies she looked at again and again during her pregnancy shaped the ways she thought about the unborn entity developing within her, even influencing her dreams about it and what it might look like. Adams goes further on to assert that '[i]t still disturbs me to realize that my conception of myself as a mother was mediated at its deepest level by obstetric technology' (1993: 270). She writes that she found the visual separation of the unborn from the maternal body in Nilsson's images served to challenge and even contradict the haptic sensations she felt from her own unborn as it moved inside her, making its presence in her body known to her and underlining the connection she felt between her body and that of her unborn. The ontology of her own pregnant embodiment, therefore, was contradicted by the images presented by Nilsson.

As I discussed in Chapter 2, obstetric ultrasound, invented and introduced into routine pregnancy care as a means of monitoring the health

DOI: 10.1057/9781137310729

and development of the unborn body for medical purposes, has become a predominantly social device for pregnant women who are undergoing these tests and their partners and other family members. Many scholars have contended that, quite apart from any medical reason for ultrasound, it has come to serve an important social function in assisting prospective parents to forge an emotional bond with their unborn (see, for example, Georges 1996, Han 2008, Harris *et al.* 2004, Mitchell 2001, Mitchell and Georges 1997, Nash 2007, Palmer 2009b, Petchesky 1987, Rapp 2000, Roberts 2012, Sandelowski 1994, Taylor 2000, 2008, Williams *et al.* 2005). For pregnant women the ultrasound images they see and preserve from their scans have become vital texts for the representation of their unborn and coming to 'know' what is growing inside them. They assist women in viewing their pregnancy as 'real' and are often part of the process of making the pregnancy public, used to prove that the unborn entity actually exists. Indeed it has been argued that in many cases, particularly when the ultrasound is taken in the first trimester or the beginning of the second, women may experience a 'technological quickening' (observing the movements of the foetus in ultrasound images) before they have felt the embodied sensations of foetal movements within them (Mitchell and Georges 1997, Nash 2007).

Ultrasound images may provide reassurance that all is progressing well with the pregnancy and that the unborn has no apparent health problems. Women who undergo IVF treatment and those who have had several miscarriages may undergo several early ultrasounds as well as regular blood tests during the first trimester of pregnancy to check on the viability of their unborn. This experience can be important for them to feel confident that their embryo or foetus exists and is developing normally (Abboud and Liamputtong 2002, McMahon *et al.* 1999, Taylor 2008). Other pregnant women who are experiencing what is considered a 'normal pregnancy' may feel anxious about undergoing an ultrasound because of the fear that something will be found to be wrong with their unborn (Harris *et al.* 2004, Lupton 1999, Williams *et al.* 2005). Where once expectant parents had to wait until the moment of birth to check that the infant had developed normally, ultrasound's visual access to the unborn allows this well before birth. Ultrasound images may therefore serve to confirm what until that moment has been a 'tentative pregnancy' (Rothman 1994), in which the couples may withhold developing an emotional bond with their unborn until they know for sure that it is normal and healthy (Mitchell and Georges 1997, Williams *et al.* 2005).

DOI: 10.1057/9781137310729

For many of the pregnant woman or couples involved the ultrasound is the first photograph of 'our baby'. Indeed prospective parents commonly refer to obstetric ultrasound as 'baby television' or 'baby movies' (Erikson 2007, Han 2008). Pregnant women may describe the ultrasound images as 'cute baby pictures' when they are showing them to others, and may keep ultrasound images in their wallets as parents commonly do with photographs of their children, even when these images depict what is only a ten-week foetus that has barely achieved a human form (Han 2008). The identification of the sex of the unborn can be a particularly important moment for prospective parents in bestowing personhood upon it (Harris *et al.* 2004, Rapp 2000). As noted in Chapter 2, sonographers contribute to this representation by using 'baby language' rather than technical language when addressing prospective parents.

Prospective fathers in particular may find it difficult to conceptualise and give meaning to what is going on in their female partners' bodies when they are pregnant. For these men the unborn may seem very mysterious and absent. For many men it is only until they see the foetus on an ultrasound or feel its movements through the wall of their partner's abdomen in the later stages of pregnancy that they begin to accept the 'reality' of its existence (Draper 2002, 2003, Han 2008, Ivry and Teman 2008, Kroløkke 2010, 2011). Indeed men often privilege the visual display of the unborn entity offered by ultrasound over the haptic indications of its presence such as foetal movement they can feel (Draper 2002). Other family members, such as grandparents, may respond to ultrasound images in similar ways, seeing them as a way of welcoming a new member into their families (Harpel and Hertzog 2010).

It is important to note that not all pregnant women and their partners respond in the same ways to sonographic images of their unborn. In my Australian research with Schmied (Lupton 1999, Schmied and Lupton 2001), while we found that the women very much appreciated the experience of 'seeing' their unborn on a sonographic image and commonly observed that this served to render the presence of their unborn as more 'real', this did not necessarily reconcile the ambiguity or uncertainty they expressed about the ontology of pregnant experience and the Self/Other distinction. It should also be acknowledged that the heightened detail of the unborn offered by 3/4D ultrasound technologies may also be challenging for some parents. Despite the lofty claims of the service providers, although 3/4D images do provide a better view of the foetal face than that offered by 2D ultrasound, the 3/4D images still appear

DOI: 10.1057/9781137310729

very different from the appearance of a newborn infant. Not surprisingly, given the close confines of the womb, the features of the unborn as they appear in 3/4D ultrasound appear squashed and distorted. One study of Jewish–Israeli men attending a 3/4D ultrasound scan with their pregnant wives found that some expressed fear and even disgust at the image of their foetus with which they were confronted. One man commented on the 'distortions in the face and the body' of the foetus and another found that the foetus looked like a 'wax figure' or a 'mummy', far from the 'cute babies' they were expecting to see (Ivry and Teman 2008).

While ultrasounds do often offer reassurance that 'all is well', this is not always the case. Viewing the unborn on an ultrasound may be a highly distressing experience when something is found to be wrong with the embryo or foetus or if women plan to electively terminate the pregnancy. The efforts of anti-abortion groups to encourage women to undergo ultrasounds were discussed in Chapter 2. Research suggests that it can be very confronting for women planning a termination to view their unborn via ultrasound imaging as they do then tend to position it as a 'baby' and may experience greater distress or ambivalence about the termination as a result (Gerber 2002). Many women attending for sonograms during the course of an apparently normally progressing pregnancy may not consider the ultrasound scan as a means of identifying medical conditions or malformations, but instead view it as a chance to see the 'baby' or as part of the routine of prenatal check-ups. The ramifications of what might happen if a foetal anomaly is discovered via ultrasound are not always explained to such women. They are therefore often unprepared if a diagnosis of abnormality in the foetus is revealed by the ultrasound (Harris *et al.* 2004, McCoyd 2007, 2009, Mitchell 2004). As an American woman who had terminated a foetus because of an anomaly noted, she was completely unprepared for bad news: 'I mean, everyone gets an ultrasound now. It doesn't seem like something you do for health things, it seems like a fun thing you do to get to see your baby' (McCoyd 2009: 512). If the ultrasound reveals that there is a problem, the personalising of the unborn that has taken place via viewing it on ultrasound can contribute significantly to women's sense of grief and anguish over making a decision about terminating the pregnancy (McCoyd 2007, 2009, Mitchell 2004). Ultrasound images in some cases also provide proof that an embryo or foetus has died *in utero*, sometimes in the absence of any other physical evidence that this has occurred. They are thus are not always the conduits of happy 'baby photos' but can rather be blunt and unexpected visual evidence of the death of an unborn entity (Keane 2009).

DOI: 10.1057/9781137310729

Research in non-western countries has demonstrated that pregnant women may use and interpret ultrasounds of their unborn somewhat differently from those in the west. For example, research on pregnant women's experiences in Vietnam (Gammeltoft 2007a, 2007b, Gammeltoft and Nguyễn 2007) found that while Vietnamese women enjoy having 'pictures of the baby', they see an ultrasound principally as a means by which they can ensure that their foetus is developing 'normally'. It is common in Vietnam for pregnant women to seek out (and pay for) numerous ultrasound scans: as many as one or more per month of pregnancy. This is because they have a high degree of anxiety about the possibility of malformation or disability in their foetus. This anxiety is underpinned by living in a society in which successful reproduction is highly valued, there are few support systems for people with disabilities, disability is highly stigmatised and in which there is high awareness of the effects of chemical warfare in past years which led to a higher rate of congenital malformations and stillbirths due to such substances as Agent Orange. Cultural concepts of unborn development also contribute to this desire to undergo constant sonograms. The Vietnamese view the unborn entity as a constantly changing fluid form in a provisional state of being. Thus, even though one ultrasound scan may show that the foetus is developing normally, this provides only short-lived reassurance. Women do not feel certain that this state of normality will continue as the foetus grows, as they see sonogram images as contingent rather than a fixed representation of the foetal being.

As this research suggests, pregnant woman and their partners do not simply or necessarily embrace ultrasound technology as the means for providing another kind of 'baby picture'. While this is an important function for many, the sociocultural and personal contexts in which these images are taken are important in bestowing meaning upon them. In some contexts the images may serve to alienate prospective parents from their unborn, contribute to feelings of ambivalence and bestow grief or anxiety or only a temporary sense of relief that the embryo or foetus is 'normal'.

Attaching/detaching: the maternal/unborn 'bond'

The issue of the emotional bond a pregnant woman has with her unborn has become a recent interest of psychology. Several psychometric scales

DOI: 10.1057/9781137310729

now exist, including the 'Maternal Fetal Attachment Scale', which seeks to measure the 'mother-fetus' relationship and determine whether pregnant women are appropriately 'attached' to or 'bonded' with their unborn (see, for example, Alhusen 2008, Van den Bergh and Simons 2009). Such scales represent the unborn as a child-like entity with whom pregnant women are expected to develop a positive emotional relationship, just as they are expected to do so once the child has been born as part of the 'normal' and 'natural' emotional responses of motherhood. To fail to develop this relationship while the child is *in utero* is viewed as potentially pathological and the possible beginning of a trajectory of emotional and mental health disorders such as post-partum depression.

Women are often highly aware of a societal expectation that they develop a close and loving 'bond' with their unborn well before birth, and may feel concerned about negative feelings they harbour towards their unborn. In my research study with Schmied (Lupton 1999, Schmied and Lupton 2001), several women articulated their concerns that they did not 'feel enough love' for their unborn 'babies' because they did not yet differentiate them from their own bodies. They were concerned about harbouring negative feelings about their unborn, worrying about how they would bond with the infants when they were born. They recognised the disjuncture between their feelings and the dominant discourse on pregnancy that represents it as a period of growing love for one's 'baby' and as a supremely rewarding and miraculous experience for women. Some of these women were also concerned that their negative feelings would somehow be transmitted to the unborn and affect them negatively. As one woman who had experienced severe nausea during several months of her pregnancy and whose pregnancy was also unplanned commented; 'I tried to blame [the nausea] on the pregnancy, not on the child, separate the two hopefully. Because I also reckon your negative emotions can be packed onto the foetus too, so I thought I'd better not' (Lupton 1999: 73).

Research on pregnancy surrogates who have been commissioned to gestate and give birth to infants on behalf of others suggests that maternal–foetal bonding is by no means inevitable and that the emotional relationship between the pregnant women and the unborn developing inside her is a product of sociocultural context. In surrogacy, indeed, such bonding is not considered appropriate, as the surrogate must relinquish the infant when it is born to the commissioning parents. For the surrogate to develop a strong affective bond with the unborn could

DOI: 10.1057/9781137310729

result in significant emotional distress on her part after birth and even in the desire to keep the infant herself. Surrogates must therefore engage in deliberate strategies to conceptualise the unborn they are gestating as Other to themselves and to maintain some emotional distance between themselves and the unborn.

A study on the concepts of motherhood among Israeli altruistic pregnancy surrogates and the women who commissioned them to gestate their children found that the surrogates worked hard at distancing themselves emotionally from the body gestating inside them. Both surrogates and the commissioning mothers constructed pregnant embodiment as separate from maternal identity. Surrogates attempted to 'feel nothing' emotionally about the unborn, to protect themselves from connection with and attachment to it. In some cases they attempted to conceptualise their bulging abdomen and its contents as separate from their bodies until delivery. In contrast, while the surrogate was harbouring their unborn, the commissioning mothers tended to view the surrogate's body as an extension of their own body, a kind of appendage, or even saw their body and that of the surrogate as one body merged together. They employed strategies, therefore, that attempted to develop a bond with their unborn even though the unborn were not physically inside them (Teman 2001, 2009).

Particularly when surrogates are paid for their services, gestating another person's unborn may become a type of physical labour. One's body may be viewed as a kind of receptacle that allows such work to take place and the unborn itself may be regarded as a valuable commodity produced through this labour. Other studies investigating the experiences of Indian surrogates who were paid to gestate infants found that these women, like the Israeli surrogates, had to work hard to distance themselves emotionally from the unborn growing within them. To achieve this emotional separation the Indian surrogates sought to position themselves only as 'carriers' or as 'prenatal babysitters' and their uteruses as separate from their 'core selves' while they were gestating others' unborn. These women said that they needed to distance themselves in such a way, to separate parts of their bodies and selves, to deal with the grief and sense of loss which they otherwise may have felt when required to relinquish the infant after birth. This research showed that Indian surrogate mothers tended to conceptualise the body growing within them as 'different from me' because its genetic makeup was that of another couple, and viewed their uterus as if it were a 'house', an

DOI: 10.1057/9781137310729

'incubator' or 'oven', temporarily providing a safe place for the 'visitor' until it reached a stage at which it was able to leave (is born) (Goslinga-Roy 2000, Hochschild 2011, Pande 2010).

While surrogacy is a far-from-usual pregnancy arrangement, the findings of these studies from very different cultural and economic contexts (Israel and India) demonstrate the ways in which the unborn body may be deliberately separated from one's Self via purposive mental strategies of symbolic detachment. Conversely a sense of attachment may be achieved even though the unborn may not be gestating inside one's own body, as in the case of the commissioning mothers in Teman's study. These women were able to construct a version of the unborn–maternal assemblage which included two maternal bodies – their own and the surrogate's – together gestating a third body – that of the unborn.

Women who are undergoing prenatal diagnostic tests often find that they similarly must work to distance themselves emotionally from their embryos or foetuses until they are shown to be 'normal' (or at least as 'normal' as the data from such tests can suggest). For many women the prospect of giving birth to an infant with serious medical or genetic problems is a deterrent to continuing the pregnancy, particularly if they have already had a child with these conditions or have family members who have experienced them (Rapp 2000, Remennick 2007, Roberts and Franklin 2004). As I observed earlier in relation to ultrasound scans, the medical evaluation of the normality of their unborn can be very emotionally challenging for pregnant women. Women are forced to confront the possibility that there may be something wrong. Those who receive false positive results suggesting that there is a problem are forced to deal with unwarranted worry. Women who receive the news that their embryo or foetus is abnormal must go through the ordeal of deciding whether or not to terminate the pregnancy (Erikson 2012, Gammeltoft and Nguyễn 2007, Gross 2010, Lupton 1999, McCoyd 2007, Rapp 1990, 2000, Rothman 1994).

Until they can be as sure as possible that their unborn is free of such conditions pregnant women may seek to emotionally detach themselves and avoid becoming too excited about their pregnancy. Some women also position their unborn as not yet 'my baby' during the first trimester of pregnancy when they know that the possibility of a miscarriage is highest, and do not inform others of the pregnancy until they feel more certain that it will be continuing successfully (Mitchell and Georges 1997). This is particularly the case for women who have experienced

DOI: 10.1057/9781137310729

repeated miscarriages (Abboud and Liamputtong 2002, Karatas *et al.* 2010) and women and couples undergoing IVF as they wait to see whether the procedure will be successful and their embryo will implant and develop normally (McMahon *et al.* 1999). Here again Rothman's (1994) term 'the tentative pregnancy' is useful to describe the view that some women have of the unborn during the time in which they await the results of tests or progress through the first trimester.

In pathologising and medicalising the emotional responses of pregnant women towards their unborn, the psychological literature on maternal–foetal bonding allows little room for the possibility that they may feel ambivalent about or even hostile towards their unborn at some stages during their pregnancy. The empirical research studies with pregnant women I reviewed earlier in this chapter suggests that these emotions may be a common and normal part of the dynamic nature of the unborn–maternal assemblage, even if they are difficult to acknowledge to others. Further, in some contexts women may feel the need to withhold developing a strong emotional bond with their unborn during pregnancy. When women are anxious about the viability or normality of the pregnancy, they are planning to terminate the pregnancy or acting as pregnancy surrogates, they may attempt to protect themselves from emotional distress by deliberately engaging in strategies that serve to distance themselves from the unborn.

Concluding comments

This chapter has gone some way towards exploring how the unborn–maternal assemblage is experienced by pregnant women. The data from interview studies across the range of social, cultural, economic and geographical contexts reviewed in this chapter demonstrate that for women themselves the lived experience of pregnancy is a shifting state involving various permutations of the unborn–maternal assemblage. This research suggests that there is nothing particularly essential or predictable about the experiences of pregnancy. The unborn may be conceptualised as mine/not-mine, part of me/separate from me, companion/antagonist, baby/parasite, Self/Other, depending on the particular context in which the woman finds herself and her own life experiences. These issues again come to the fore in cases of pregnancy termination, pregnancy loss and choices about the use of surplus embryos from IVF treatment, as I go on to show in Chapter 4.

DOI: 10.1057/9781137310729

4
Death, Disposal and the Unborn

▶

Abstract: *This chapter addresses the topics of abortion, pregnancy loss and the disposal of aborted embryos and foetuses and surplus IVF embryos. While these are very different issues, they all refer to the ways in which the unborn are understood in relation to the continuum of 'life' or 'personhood' and to concepts of and practices around the death or loss of vitality of the unborn. Often very highly emotional debates have been provoked in relation to how best to regulate and respond to these interventions and practices. These debates highlight the ways in which political, historical, moral, ethical and religious perspectives shape notions of the unborn. This chapter includes discussion of abortion legislation in various countries, women's experiences of elective termination of pregnancy, how women and couples make decisions about the disposal of aborted embryos and foetuses or IVF embryos, how bioscientific workers approach using the unborn in their work and the mourning and memorialisation of the dead unborn.*

Key words: abortion; pregnancy loss; IVF embryos; women's experiences; legislation; disposal choices

Lupton, Deborah. *The Social Worlds of the Unborn.* Basingstoke: Palgrave Macmillan, 2013. DOI: 10.1057/9781137310729.

DOI: 10.1057/9781137310729

Cultural variants in attitudes towards abortion

The longstanding debate over the morality and legality of the elective termination of pregnancy has brought to the fore contestations over definitions of the unborn for centuries. Abortion is a very common practice globally: it has been estimated that the majority of women worldwide undergo an abortion at some stage during their lives, either legal or illicit (Kumar *et al.* 2009). Despite the frequency of the experience, significant social stigma is attached to women seeking abortions in many countries and the practice is legally prohibited across a range of countries, both developing and developed. Even in those countries in which abortion is made readily and legally available, women may still experience abortion-related stigma because the practice contravenes religious beliefs or norms concerning the desirability of motherhood, appropriate sexual behaviour for women and the moral status of the unborn. They therefore may choose not to disclose to others that they have terminated a pregnancy (Kumar *et al.* 2009, Shellenberg *et al.* 2011).

According to the Center for Reproductive Rights, which maintains a digital map showing current abortion laws by country, over 60 per cent of the world's population reside in countries that have permissive abortion laws, while 25 per cent live in countries in which abortion is prohibited entirely or only to save the pregnant woman's life. A further 14 per cent live in countries where abortion is permitted only to save a woman's life or in the interests of her health. The most permissive abortion legislation is found primarily in countries in the northern hemisphere (North America, Europe and northern Asia) (*The World's Abortion Laws* 2013). Most developed countries allow for early-term abortions (performed in the first trimester of gestation). However there is little agreement on the more contentious issue of late-term abortions because the foetus is viewed as more human-like the closer it edges towards the age of viability. In those countries that allow abortion, definitions of personhood, and therefore the stage of unborn development at which abortion is allowed, vary.

As discussed in Chapter 1, concepts of when life begins vary across cultures and societies. These are pivotal in shaping attitudes and policies related to abortion (Morgan 1997, Rylko-Bauer 1996). As Gross (2002: 203) has pointed out, 'The same moderately malformed 25 week old fetus might be aborted in Israel, delivered but not necessarily resuscitated in Denmark, resuscitated but not always treated aggressively in the UK and treated aggressively in the US.'

DOI: 10.1057/9781137310729

The doctrines of the various major world religions each teach a different perspective on the unborn. In Judaism the unborn is considered a person in its own right only once its head has emerged from the mother's body and it has drawn its first breath. The Roman Catholic Church's official position is that human life and personhood begin at the moment of conception. Hinduism proposes that human life does not have a clear beginning or clear end (Dubow 2011). In traditional Islamic teaching the unborn entity becomes a person at 'ensoulment', occurring either at 40, 90 or 120 days after conception, depending on the school of thought. Abortion is considered acceptable if it takes place before this time as long as there is a justifiable reason, but is generally forbidden after the point at which 'ensoulment' is considered to have occurred (Hessini 2007). Buddhist teachings view all killing as wrong and position the unborn as living beings from the moment of conception (Perrett 2000).

Religious doctrines on the unborn are frequently phrased through other cultural understandings and practices, and therefore are not always adhered to strictly. While, for example, Catholicism is the dominant religion in Ecuador, abortion is viewed as self-mutilation rather than killing a unborn entity because the embryo/foetus is not considered an individual organism, separate from the maternal body (Morgan 1997). Buddhist thought, particularly as it is practised in East Asian countries, tends to confront a dilemma because it commits to a compassionate view of both the pregnant woman and the unborn entity. Despite the condemnation of abortion in Buddhist teachings, the abortion rates in countries such as Korea, Thailand, Vietnam, Taiwan and Japan are high compared to western countries because abortion is considered more acceptable a form of contraception than the contraceptive pill. Japanese Buddhists, for example, view abortion as a regrettable but necessary practice, and represent the terminated pregnancy as a 'child' that due to circumstances must be 'returned' to the world of the non-living (Perrett 2000).

Many formerly communist or socialist states have promoted abortion as an appropriate means of controlling family size and population growth. In China the unborn are considered to be integrated with the maternal body rather than entities in their own right. There is consequently little moral or ethical debate about the appropriateness of abortion (Crow 2010, Rigdon 1996). In that country abortion has been an important part of instrumental state-level policies concerning population control. It is regarded as a vital secondary measure of fertility control to be used when contraception fails, and is available at any stage of gestation. Pregnant

DOI: 10.1057/9781137310729

women in China have sometimes been coerced by government officials to undergo an abortion in the interests of maintaining the government's one-child policy (Rigdon 1996).

Since the 1980s, China has attempted to promote other forms of contraception to its citizens, resulting in abortion levels falling (Zheng *et al.* 2013). Nonetheless, because sons are considered far more desirable than daughters in Chinese culture, particularly in the context of strict fertility restrictions, since obstetric ultrasound became widely used in China in the late 1970s the abortion of female foetuses has become a common practice. Although abortion for sex-selection reasons is now illegal, it still occurs on a widespread basis and has resulted in a significant gender imbalance in that country (Crow 2010). Abortion is also widely accepted as a form of contraception and population control in India. As in China, sons are much more favoured than daughters, and many Indians have used technologies such as amniocentesis and ultrasound to determine the sex of foetuses and to selectively abort those that are found to be female (John 2011). Abortion for sex-selection purposes in favour of male foetuses is also increasingly found in developed countries such as Canada and the UK with high numbers of immigrants from countries such as India, China, Vietnam and Korea (Vogel 2012).

The legal systems and religious beliefs of France, Italy, Portugal, the Netherlands and the UK influenced abortion legislation in their former colonies in Africa and the Middle East. While the abortion legislation in those European countries may have changed since colonial times, this has often not been reflected in the statutes of their former colonies, many of which retain highly restrictive abortion laws (Brookman-Amissah 2012, Hessini 2007). Among the countries of Northern Africa, sub-Saharan Africa and the Middle East, only South Africa and Zambia have liberal abortion laws. In such countries as Iran, Iraq, Syria, Afghanistan, Egypt, Tanzania, Anglo and Libya abortion is either prohibited outright or allowed only to save the pregnant woman's life or in cases of rape or foetal disability. Several other countries in this region, including Saudi Arabia, Algeria, Morocco, Ethiopia, Namibia, Zimbabwe and Kenya are somewhat more permissive, permitting termination of pregnancy in cases where the pregnancy threatens the woman's health (*The World's Abortion Laws* 2013). As a result of these restrictions as well as lack of access to professional health care providers, many women in African and Middle Eastern countries are forced to attempt unsafe self-procured abortions or engage untrained lay abortion practitioners, resulting in a

DOI: 10.1057/9781137310729

high burden of ill-health, disability and increased risk of early mortality (Brookman-Amissah 2012, Vlassoff *et al.* 2009).

In Anglophone countries such as the UK, the USA, Australia and Canada abortion is legal in certain circumstances, sometimes involving the approval of one or more medical practitioners and contingent on gestational age. In these countries concepts of autonomy, individuality and liberal humanism shape responses to both the pregnant woman's right to control her own body and the unborn organism's right to live. In the USA, the landmark *Roe v. Wade* decision by the American Supreme Court in 1973 legalised abortion up to the 'age of viability' (at that stage, defined as 28 weeks but possibly as low as 24 weeks' gestation). However the abortion issue remains contentious in that country. Anti-abortion activists are particularly militant, some even using violence in targeting abortion clinics and those who work there, and conservative Christian politics has some influence in government decision-making on abortion policies (Hadley 1994, Parkes 1999). The foetus in the USA has consequently acquired a measure of legal personhood and protection from the age of viability which is far greater than in most other developed countries. It is very difficult to seek permission to terminate after the age of viability, even if the foetus is found to be grossly malformed or have a life-limiting medical condition (Gross 2002).

Political moves to redefine their status and renewed efforts from anti-abortion spokespeople and groups have recently emerged in the USA. Indeed it has been argued that in many US states elective abortions are harder to access now than at any time since the *Roe v. Wade* decision 40 years ago because of the increased dominance of conservative activists and politicians in that country (Pickert 2013). Since 2008, what have been termed 'personhood' initiatives have emerged in the USA, gaining some legitimacy and forcing legislative battles in several states (Paltrow and Flavin 2013). These initiatives, many of which have been instigated by an overtly Christian organisation entitled 'Personhood USA', are centred on attempting to define personhood as beginning at the moment of fertilisation. If the legislative changes proposed by 'Personhood USA' in various US states were to be brought into being, IVF personnel and embryologists who freeze or dispose of surplus embryos and scientists who use hESCs, as well as medical practitioners who perform abortions or even remove ectopic pregnancies would potentially be open to criminal prosecution for threatening the life of embryos or foetuses (Collins and Crockin 2012).

DOI: 10.1057/9781137310729

In eastern European countries, abortion has been used as a major form of fertility control, actively promoted by state-run medical bodies due to the expense and unavailability of reliable contraception (Parkes 1999, Rylko-Bauer 1996). Russia, Romania, Estonia, Ukraine, Bulgaria and Latvia lead the world in abortion rates per head of population, with Russia in particular far ahead of the rest of the world. In that country, surveys of attitudes among the population suggest that the majority of Russians view abortion as an acceptable form of fertility control. Indeed the Soviet Union was the first country to legalise abortion on demand, in 1920 (Solodnikov 2010).

Several countries with a dominant Roman Catholic heritage, including Ireland and Latin American countries such as Chile, Brazil, Venezuela, Paraguay and Mexico, have criminalised termination of pregnancy, adhering to the Vatican's position (*The World's Abortion Laws* 2013). Thus, for example, Irish legislation valorises protection of the life of the unborn and positions these entities as having equal rights to those of the pregnant woman. Abortion is permitted in Ireland only if the life (not simply the health) of the pregnant woman is threatened by continuing with the pregnancy. Even in this case, abortion is not acknowledged publicly in Ireland as an acceptable choice for women. Doctors are often concerned about being prosecuted if they carry out an abortion and the courts decide that the woman's life was not sufficiently threatened. As a result many Irish women travel to Europe to procure pregnancy terminations (Kay 2012, Lalor *et al.* 2009).

The case in November 2012 of the death in Ireland of Savita Halappanavar, a 31-year-old Indian immigrant who sought hospital treatment for a miscarriage and died three days later from septicaemia, drew worldwide attention to Ireland's legal and cultural stance on abortion. Halappanavar's husband claimed that despite her request for a termination she was told that doctors could not do so because the foetus still had a heartbeat. The delay apparently caused the systemic blood infection that led to her death. Critics contended that the hospital's focus on the foetus rather than on the medical condition of the pregnant woman meant that Halappanavar did not receive appropriate treatment in time. Pro-abortion advocates argued for an urgent revision of Irish abortion legislation in the wake of this case, arguing that Irish abortion law is in contravention of the European Convention of Human Rights (Kay 2012).

Several other European countries, despite their predominantly Catholic religious heritage, are comparatively far more liberal about

DOI: 10.1057/9781137310729

abortion: indeed with the exception of Poland and Germany, European countries have among the most permissive abortion legislation in the world (*The World's Abortion Laws* 2013). The French, for example, are avidly pro-abortion and do not tend to place as much symbolic importance upon the embryo or foetus as do Anglophone countries. Indeed French people tend to use the word 'egg' rather than 'embryo' or 'foetus' to describe the unborn organism at its early stages of development, thus not overtly acknowledging that an entity with a unique genetic makeup has been conceived, far less a 'baby' (Gerber 2002). In Denmark too, abortion is relatively uncontroversial and is available on demand. It is considered more important that women should want their offspring so that they can be responsible mothers and produce productive citizens than the life of the unborn be preserved (Hoeyer *et al.* 2009).

In neighbouring Germany, however, termination of pregnancy is technically illegal, although it is within the discretion of parliament not to punish those who seek abortion if undertaken in the first trimester and if the pregnant woman undergoes state-regulated counselling (Diepper 2012). According to the German constitution the embryo is viewed as human from the time of implantation into the uterus. The German position has been influenced both by Catholic religious beliefs and the legacy of the Holocaust. Concern about the Nazis' unethical and often inhumane treatment of research subjects from Jewish and other minority groups, including people with disabilities, has produced a heightened sensitivity to decisions over the value of human life in that country (Hashiloni-Dolev and Shkedi 2007, Hashiloni-Dolev and Weiner 2008, Krones *et al.* 2006, Sperling 2008).

While, in contrast, Israel is a country dominated by Jewish religious belief and a strongly pronatalist ethic, Israelis do not consider the moral status of the unborn independently of their relations with their families. They tend to deploy a relational ethics, viewing the unborn entity as an organic part of the mother rather than as individuated from her. It is thus not viewed as having its own rights. When health considerations are at stake, the pregnant woman's interests are seen to override those of the embryo/foetus (Hashiloni-Dolev and Shkedi 2007, Hashiloni-Dolev and Weiner 2008). Israeli law therefore allows termination of the unborn at any stage of gestation. Once the infant is born, however, it acquires full moral personhood (Gross 2002, Hashiloni-Dolev and Weiner 2008).

As I observed in Chapter 1, regardless of their geographical location, opponents of abortion tend to adopt the 'personhood begins at

DOI: 10.1057/9781137310729

conception' argument, often based in religious belief, and use this to support their opposition to pregnancy termination, positioning it as 'murder' of an 'unborn child'. Pro-abortion advocates avoid any kind of religious argument for one based on a rights-based position. They attempt to position the unborn entity as not yet fully a moral person, and as therefore secondary in importance to the needs and rights of the woman in whose body it is located. From this perspective termination of a 'pregnancy' is therefore not viewed as the 'murder' of a 'person'.

The majority of feminist writers adopt this position, supporting abortion as a means for women to control their bodies and fertility and thus to be able to enjoy full autonomy and equality of opportunity. From their perspective, the opportunity to access abortion, regardless of the reason for doing so, is viewed as a fundamental human right (Avalos 1999, Boucher 2004, Ginsburg 1990, Hartouni 1991, 1992, Hopkins *et al.* 2005, Morgan 1996, 2009). However many feminists are much less supportive of abortion practice when it is used as a technique to select female foetuses and abort them, regarding this as a form of misogyny (John 2011). Some feminists are also highly aware of the complexities of pro-abortion discourse when it is employed to support eugenic and racist policies. They have questioned whether the concept of 'private autonomy' that is so frequently championed in feminist pro-abortion discourses is as free as such advocates suggest of social, cultural, economic and political constraints (Sharp and Earle 2002, Smyth 2002).

Women's experiences of elective abortion

I noted earlier in this chapter that abortion stigma is a common feature of many countries regardless of their legal and cultural position on elective abortion. A related dominant discourse, particularly espoused in anti-abortion rhetoric, is that women who undergo abortions will experience long-lasting regret and emotional distress (Greasley 2012). However the academic research on this topic suggests that most women who undergo elective abortions in the first trimester of pregnancy because the pregnancy is unwanted do not articulate such responses (Major *et al.* 2009, Major *et al.* 2000). A study of attitudes to and experiences of abortion in five countries (Pakistan, Peru, Mexico, the USA and Nigeria) found that while they may speak of feelings of shame, guilt and sadness, and for those who are religious, of committing a sin, women tend not to regret terminating a

DOI: 10.1057/9781137310729

pregnancy because they felt it was the right thing to do for themselves and their families in their circumstances (Shellenberg *et al.* 2011).

Women who have undergone a pregnancy termination demonstrate a range of perspectives on the 'personhood' of the unborn. One American study found that some conceptualise this experience as the 'loss of a baby', albeit an unwanted one, and feel deep grief at this loss. Others, in contrast, do not see the unborn as a 'baby', but rather as a burdensome 'pregnancy'. Still others continue to imagine the child they may have had for decades, constructing a possible life or appearance for them (Avalos 1999). Similar findings were evident in more recent research involving Australian women considering or undergoing an abortion (Kirkman *et al.* 2011). The notion that they were 'killing my own child' or 'getting rid of a child' was evident in some of these women's accounts. Those who already have children sometimes found contemplating an abortion difficult because they thought of their current children and their love for them and imagined the unborn as growing up to be a sibling for these children. However other women expressed nothing but relief that their worries and anxieties about having an unwanted child would be assuaged by termination, or represented the unborn as 'just a cluster of cells' or 'not yet a living thing'.

Social concepts about when it is the 'right time' to have a child and which kinds of parents are 'appropriate' influence societal attitudes to pregnancy termination. Adolescent girls who become pregnant find themselves in a difficult position. Having children very young is often considered wrong, disadvantaging both the child who may be born and also the young mother, but so too is terminating a pregnancy. A pregnant teenage girl is therefore the subject to moral opprobrium whichever choice she makes. A report of three English studies found that young women who found themselves in this predicament again expressed a range of views, from relief that they could access abortion because they were 'not ready' to be a mother, to distress, shame and guilt about 'killing a baby' (Hoggart 2012).

As noted in focus groups held with British women across a range of ages, including those who had had an abortion, a wanted pregnancy tends to be viewed as a 'baby' at whatever stage of gestation it has reached (Pfeffer 2008). Those participants who had borne children tended to use the term 'embryo', 'foetus' and 'baby' as if they were interchangeable, and noted that they felt differently about abortion now that they had given birth to and cared for children. The women who had undergone a

DOI: 10.1057/9781137310729

pregnancy termination still sometimes felt a sense of ownership over the aborted embryo or foetus. Some were subsequently reluctant to entertain the idea of donating their aborted embryos or foetuses to hESC research. The women saw embryos and foetuses as having a different moral status from a body organ, therefore, and positioned them as different from their own bodies, as having their own unique status. As one woman put it, 'it's not an organ of your body, it doesn't belong to you, it's growing inside of you' (Pfeffer 2008: 2552).

Some of the participants in Pfeffer's study expressed discomfort and horror at what might happen to the terminated embryos or foetuses at the end of a working day in an abortion clinic. It was notable that several of these women made a distinction between how embryos or foetuses from spontaneous miscarriages or abortions for medical reasons should be treated compared to those who were terminated because of an unwanted pregnancy, noting that the former but not the latter should be given some sort of rite to mark their death (Pfeffer 2008).

Clearly, the reasons why women seek abortions may influence their views of the entities that are destroyed via the process. Not all pregnancies are terminated because the woman does not want to bear a child at that stage in her life. As discussed in Chapter 3, some women undergo what have been termed 'selective abortions' following medical advice that their unborn has major health or developmental problems. Conditions such as Down's syndrome, muscular dystrophy, thalassemia, haemophilia, cystic fibrosis, as well as structural malformations such as anencephaly, cleft palate, congenital heart defects and spina bifida, can now be diagnosed prenatally using testing and screening procedures. Women who have been told following prenatal screening or testing that a wanted foetus has a genetic or other health condition or anomaly are forced to confront a decision about whether or not to terminate the foetus. They then must deal with their grief and possible feelings of shame and guilt if they decide to do so (Gammeltoft 2007a, Gammeltoft and Nguyễn 2007, McCoyd 2007, 2009, Rapp 2000).

As an American study interviewing women in this situation found, they must struggle with the competing societal expectations that they 'bond' with and love their unborn from when they are first aware that they are pregnant and also that they should not deliver an infant with a serious health condition or malformation. The stigma of abortion also may affect these women, who are highly aware that their society tends to condemn pregnancy termination, and some decide to keep their

DOI: 10.1057/9781137310729

experience secret as a result. They are also aware of the negative socio-cultural implications of becoming parents to a child with significant disabilities and the impact of such an event on their lives: hence their decision to terminate the pregnancy. Yet many of these women had viewed their unborn on an ultrasound, discovered its sex and even had already given it a name, thus rendering it more infant-like and more a part of the family (McCoyd 2007, 2009). As one woman said of the foetus she aborted, 'Jared was a living, breathing little baby to us, in life and death ... We thought of him as our son, our first child, and that he'll always be' (quoted in McCoyd 2007: 42).

Decisions about disposal

Since the development of IVF, the argument over to what extent embryos should be considered as persons has extended into public consideration of the ways in which *ex vivo* embryos that are surplus to requirements should be treated. Hundreds of thousands of such embryos around the world have been created for the purposes of treating infertility and have been cryogenically preserved, either because the individuals or couples who commissioned their creation wish to keep them for future attempts at conception or because they cannot decide what to do with them. The existence of these embryos raises many moral and ethical issues for both the people whose genetic material was used to create the embryos and for the workers who handle them.

In several jurisdictions, time limits are placed on how long the embryos may be cryogenically stored. This time period varies from one year in Austria and Denmark, to five years in the UK, Iceland, France, Belgium and Croatia, to ten years in Finland, Israel and Spain. Some countries, like Australia, have legislated on a state-by-state basis, some countries, like the US, do not enforce any time limits and others, such as Germany, do not allow cryopreservation of embryos (*Embryo Bank* 2012). In those countries where there are time limits, at the expiry of the designated time period couples or individuals are forced to make a decision about what to do with these embryos or else leave it to the clinic to dispose of the embryos. These issues have intensified in the wake of hESC, because these embryos offer ideal material for use in creating stem cell lines for therapeutic purposes. Couples who have created the embryos now may be asked whether they wish to donate them to hESC research, to other

DOI: 10.1057/9781137310729

infertile individuals or couples or agree to destroy them if they do not wish to use them for reproductive purposes.

Some IVF clinics also offer what is termed 'compassionate transfer' for unwanted embryos, in which the embryo is transferred to the woman's body in a location (her cervix or vagina rather than her uterus) or time period (at menopause or when the uterine wall is not prepared for implantation) in which it is certain it will not develop and therefore 'dies' in her body. This practice is viewed by the clinics that offer it and the individuals who take it up as more ethical and kinder than disposing of the embryo as if it were a type of medical waste. It is also seen as closer to a 'natural' embodied process such as menstruation or miscarriage and thus as transforming the artificially created *ex vivo* embryo into a more natural entity that is closer to a potential baby than to clinical waste matter. The practice of compassionate transfer is also a means of providing a ritual of disposal that marks the loss of the embryo while at the same time recognising that this process is not exactly a 'death' as we commonly understand the death of a person (Ellison and Karpin 2011).

In discourses concerning how best to deal with surplus IVF embryos, these entities are often portrayed as infants or children who are at risk of murder by the scientists who created them. Anti-abortionists have frequently represented spare embryos destined for discarding as imperilled infants in need of 'rescue'. For instance, in a debate over disposal of their embryos between an American couple who had separated, the woman, who wanted to use the embryos to bear a child, described them as her 'unborn children' who she wanted to rescue from the 'concentration camp of the freezer' (Hartouni 1991: 50). The discourse of embryo 'orphans' and embryo 'adoption' also configures embryos as already infants or children. One Christian adoption organisation in the USA has three arms: domestic adoption, international adoption (both of infants or children) and embryo adoption. The embryo adoption program is named the 'Snowflakes Frozen Embryo Adoption Program', because, according to the organisation's website, like snowflakes, each embryo is unique, and by inference, special. The website describes frozen embryos as 'preborn children' who are 'waiting for a loving home' (*Snowflakes Embryo Adoption* 2012).

The embryo adoption program was formed in the late 1990s in the wake of public attention being drawn to the increasing numbers of embryos being produced in IVF programs surplus to requirements which eventually had to be disposed of by clinics (estimated to be at

DOI: 10.1057/9781137310729

least 600,000 in the US alone). Such 'massacres' caused consternation among some people associated with this adoption organisation, who subsequently established the embryo adoption program (Ganchoff 2004, Langstrup and Sommerlund 2008). When President George W. Bush made the decision to withdraw US federal funding from hESC research in 2001, he was shown on television programs surrounded by some of the 'Snowflake children' who had resulted from 'embryo adoption', many wearing t-shirts with the legend 'Former Embryo' upon them (Langstrup and Sommerlund 2008).

If embryos are bestowed with the ontological status of a 'baby' or 'potential baby', the decision to discard them or donate them is rendered very difficult for the couples who have contributed their gametes to create these organisms. Interviews with couples in the UK who had embryos awaiting implantation, or who were debating about what to do with their spare embryos, found that their deliberations invariably began with 'baby talk' or using the concept of 'our baby' (Haimes *et al.* 2008). So too, an Australian study of couples undergoing IVF found that some of the participants described donating their spare embryos to another couple as akin to allowing another couple adopt their 'children' and they were therefore reluctant to do so (Waldby and Carroll 2012). Japanese couples undergoing IVF similarly described their spare or failed embryos as their 'lost children'. Those who had made the decision not to donate their spare embryos to research saw such an act as agreeing to kill their children equivalents and cutting their bodies into pieces. The elision of embryo with infant/child was particularly evident in the words of those couples who had not yet succeeded in producing a live child: for them, the embryos were viewed as their only children (Kato and Sleeboom-Faulkner 2011). A Scottish study produced similar findings (Parry 2006).

In the context of using PGD of embryos to avoid possible medical conditions, the notion of discarding defective embryos may be particularly traumatic for couples who have previous children with the same genetic disorder or experienced the death of such affected children, and who therefore see the embryos as similar to their other children, requiring similar compassion and dignity (Karatas *et al.* 2010). By contrast, in India, where few frameworks or legislation exist to govern what is a rapidly growing market in IVF and surrogacy for both resident Indians and reproductive tourists, the idea of donating surplus embryos to other infertile couples or medical research is generally supported by

DOI: 10.1057/9781137310729

local couples undergoing IVF. Embryos do not tend to be represented as 'babies' in Indian culture. There is therefore little religious opposition to abortion or the use of human embryos in stem cell research or debates on the moral status of the embryo (Gupta 2011).

Another Australian study (de Lacey 2007) found that couples tended to employ several explanatory systems when defending their decisions about how to dispose of their surplus embryos. All the couples acknowledged the potential of their embryos to become children. For some couples, donating their embryos to others was chosen because they saw the decision to donate them to other couples as a way of realising this potential. They were able to donate because they considered the act of gestating a foetus, and not only genetic relatedness, as an integral dimension of parenthood. These couples also tended to consider the embryo as a conglomeration of cells rather than as a baby. Other couples, however, found the concept of other people raising their children as problematic, as they considered their genetic relationship to the embryo as implying that it was part of their own family. These couples therefore chose not to follow this course of action. They considered the embryo as one of their own children whose development was suspended, a 'virtual child'. The interviewees in this study also often referred to the discarding of embryos as similar to a pregnancy termination.

Part of the value ascribed to the act of donation of one's embryos to other couples or scientific research is the idea that one is benefiting others selflessly. Research from a range of developed countries, including the UK, the USA, Australia, Switzerland, Japan and Denmark, has shown that people considering what to do with their surplus IVF embryos often bestow value upon their embryos based on such factors as the expense, emotional investment and trouble involved in their creation, their intrinsic worth as unique human tissue and how they could help others with infertility problems or health conditions. If they decide to donate the embryos, this act provides many such individuals with the sense that they are not going to waste. It serves to render the embryos more precious and bestows them with significance, and gives the process of undergoing IVF meaning, even if no viable pregnancy resulted from the creation of the embryos (de Lacey 2007, Franklin 2006b, Haimes *et al.* 2008, Kato and Sleeboom-Faulkner 2011, Parry 2006, Svendsen 2007, Waldby and Carroll 2012, Waldby and Cooper 2010). These meanings are frequently conveyed in medical professionals' discussions with couples or individuals. A Danish study (Svendsen and Koch 2008) and

DOI: 10.1057/9781137310729

British research (Ehrich *et al.* 2010, Wainwright *et al.* 2006, Wainwright *et al.* 2008, Williams *et al.* 2008) found that *ex vivo* embryos which are surplus or low quality and those *in vivo* embryos or foetuses which have been aborted are typically reconfigured in professionals' discourse from waste matter to potential valuable research or therapeutic material.

Bioscientific research and definitions of the unborn

Legislative bodies have recently been forced to grapple with the thorny issue of defining embryonic life as part of issuing guidelines and regulations about how embryos should be created and dealt with in the context of hESC. Discussions over the use of surplus frozen IVF embryos in hESC research and regenerative medicine have often centred on details concerning the stage of development of the embryo and its potential for future development. As is the case with abortion, different countries hold differing approaches to the morality of the use of embryos in hESC research and therapeutics based on their own history and the relative dominance of religious belief. While, for example, the UK, Denmark and Australia allow surplus IVF embryos to be used for hESC research, under the Embryo Protection Act Germany maintains a total ban on the use of German embryos in hESC science, although it permits the use of hESCs from other countries. Consonant with Germany's position on abortion, in that country the embryo is entitled to legal protection from the moment of fertilisation. Any practice that involves its destruction or potential destruction, such as cryogenic preservation, PGD and use of domestic embryos in hESC, is therefore prohibited (Hashiloni-Dolev and Shkedi 2007, Krones *et al.* 2006).

In the Netherlands, a distinction has been made between embryos created and used for IVF purposes, and therefore considered to be 'humans in the making', having the intrinsic potential to develop into a foetus when implanted, and those embryos that are spare or created purely for research or therapeutic purposes, which will never have this possibility and thus are treated as non-human (Kirejczyk 2008). In Italy the genetic makeup of embryos has been emphasised in debates about hESC science. The embryo was described by opponents of hESC science as a tiny citizen, 'one of us' (full human subjects) because it had its own unique genome and because 'we have all been embryos'. To freeze embryos and use them for scientific purposes was represented as

DOI: 10.1057/9781137310729

the same as subjecting fully human subjects – 'us' – to these procedures (Metzler 2007: 421). Because of this stance, Italy, like Germany, forbids embryos produced within its borders to be used for stem cell research and also prohibits the donation or even the production of spare embryos from IVF. It does allow the importing of stem cells from other countries (Metzler 2007).

The Human Fertilization and Embryology Authority in the UK and the European Society of Human Reproduction and Embryology Task Force on Ethics and Law have ruled that while it has a unique genetic makeup, the pre-14-day embryo – technically termed the 'pre-embryo' and also referred to as the blastocyst – lacks distinct individual moral status. These authorities therefore deem it acceptable to discard it or use it for scientific research or therapeutic procedures. In contrast the post-14-day embryo is positioned as human and therefore as not appropriate to be used for hESC purposes (Ehrich *et al.* 2008, Rubin 2008).

In the USA the legislation concerning hESC research is rather complex. The Bush administration banned US federal funding for hESC research in 2001, but this was overturned by the Obama administration in 2009. Some but not all of the US states have also provided funding for hESC. Various legal challenges have since been made to the current legislation and disputes continue in the courts (Johnston 2010). Australian legislation allows for the use of surplus IVF embryos to be used in stem cell research, regardless of stage of development, but proscribes the creation of embryos solely for research unless they are created using nuclear transfer technology (therapeutic cloning) rather than fertilisation using human sperm and ova (*Use of Human Embryos in Research* 2009).

In discourses in the medical scientific literature on embryonic or foetal tissue research, the unborn body is portrayed as a disembodied tool for research or a therapeutic technology. In the laboratory the terminology adopted for discussing these bodies breaks them down into their component parts, such as tissues, organs and cells. They thereby become work objects which must be dehumanised and reconstructed as 'matter' so that they can be dealt with dispassionately as part of medical or scientific work (Casper 1998, Pfeffer 2009). This is partly as a means of traversing the ethical and political minefield of abortion politics, as it allows scientists to represent embryonic and foetal tissue as non-human in a context in which anti-abortion activists call for the recognition of the unborn as human and argue against using tissue derived from induced abortions. It also helps those working with such

DOI: 10.1057/9781137310729

tissue to overcome their possible emotional reactions to working with the remains of unborn bodies (Ganchoff 2004).

However they are obtained, embryos and foetuses used in science undergo a process of decorporealisation that allows the scientists to work upon them. For example the woman who has undergone a pregnancy termination is described as a 'donor' and her 'unwanted pregnancy' is viewed as biological matter, like a donated body organ or blood, which eventually is transformed into stem cell material. The requirements of the laboratory structure the way in which the unborn entity is treated once removed from the uterus and sometimes even the method of abortion itself. Embryonic tissue becomes matter to be cleansed, purified and stored, which reconfigures it into a biological research tool. Scientists working on this material must sometimes overcome their emotional response to being able to discern body parts that are recognisably human and 'baby-like'. Once this material is rendered into cells, this link with personhood is erased (Ehrich *et al.* 2010, Pfeffer 2009).

For many professionals in IVF and hESC science, embryonic bodies are positioned as 'moral work objects': that is, objects with which these workers interact and use as part of their working lives but in a context in which moral values and judgements are to the fore (Ehrich *et al.* 2008, Ehrich *et al.* 2010, Ehrich *et al.* 2007, Wainwright *et al.* 2006, Williams *et al.* 2008). Within this category of 'moral work objects' are a range of definitions based, for example, on the age and stage of development of the embryo and the reason why it was first created. At the blastocyst stage the embryo tends to be considered by some workers as definitely non-human and more as organic material. To discard these embryos because they have been found to be abnormal or surplus to requirements, therefore, does not pose a moral problem, for they are seen as having no human attributes such as pain sensation or consciousness, and are located in an environment (the laboratory) where they cannot grow into an infant (Ehrich *et al.* 2008, Wainwright *et al.* 2006).

For other workers, however, these embryos are viewed as special cells with potentiality to be human: 'potential people'. Some go even further, defining these embryos as 'a life', 'beings' and even 'babies'. For these workers, discarding embryos, although part of the expectation and routines of their work environment, can be more problematic and replete with ambivalence. They define discarding embryos as the 'death' of something and as a 'painful' experience and must sometimes force themselves to be emotionally distant from the embryos and not to think

DOI: 10.1057/9781137310729

too much about the ethical implications of their work (Ehrich *et al.* 2008, Wainwright *et al.* 2006).

Some writers have identified a third position, somewhere between the two polar and irreconcilable extremes of 'embryo as human' and 'embryo as non-human', which receives particular expression in debates on the use of embryos for scientific or medical purposes. From this perspective the embryo is viewed as a unique type of human tissue which, because of its potential to be human is precious and unique and deserves special treatment compared with other biological matter, but is not given the moral or ethical status of 'fully human' and thus may legitimately be used in biomedical work. This position has been articulated in the legislation and guidelines on embryo use of several countries, in which it is explicitly noted that embryos should be handled with 'respect' in the laboratory context and should not be created or destroyed frivolously or unnecessarily (Jackson 2006, Steinbock 2000). It is interesting to note that this 'respect' appears not to be a concern in relation to aborted embryos or foetuses, which are frequently designated as without any kind of special value, as just another type of 'clinical waste' once extracted from the uterus, and are disposed of in an incinerator without ceremony (Pfeffer 2009). If these embryos are donated to hESC science, however, they then acquire a different kind of value as a precious research resource and thus as worthy of the 'respect' accorded to surplus IVF embryos (Kent 2008).

Mourning and memorialising unborn death

As discussed in Chapter 2, the spontaneous loss of an embryo or foetus once a pregnancy has been confirmed, and in some cases, the elective termination of a pregnancy, is often experienced by women or couples as the death of a child, even if this occurs in the very early weeks of pregnancy. Research on experiences of pregnancy loss has found that when it happens in the early stages, people often find it difficult to reconcile their feelings of grief with a dominant view expressed by others that no real death has occurred. The woman or couple as well as their family members may grieve for the 'proto-person' they had already begun to anticipate and plan for as part of their families. The grief they may feel is for the loss of this baby, for whom they have imagined a future, a personality, an identity, to whom they may have even already given a name (Abboud and Liamputtong 2002, Davidson 2011, Frost *et al.* 2007,

DOI: 10.1057/9781137310729

Jonas-Simpson and McMahon 2005, Kevin 2011, Layne 2000, 2003a, 2003b, Letherby 1993, Littlewood 1999, McCoyd 2007, McCreight 2004).

This experience may be particularly difficult for women who have undergone IVF and thus have had to endure numerous medical procedures and a greater intensity of focus around the fertilisation and implantation of embryos and monitoring of their progress after having already experienced the disappointments of infertility. Once an embryo has been implanted, hope is aroused that it will continue to develop until pregnancy is confirmed. When the embryo does not survive women and their partners can be devastated (Kevin 2011). As Kevin notes, in the context of assisted reproduction treatment, 'the definition of pregnancy in pregnancy loss stretches to accommodate the dashed hopes that can occur anywhere along a continuum from egg collection to stillbirth' (2011: 856). So too, pregnancy loss support material such as that presented on the official websites of organisations established for this purpose tend to represent the 'loss of a baby' as occurring any time from conception to birth. Particularly in the USA this material sometimes uses very similar imagery and rhetoric to anti-abortion organisations (Kevin 2011).

Miscarriage is often a difficult topic for people to discuss and acknowledge, and as a result, miscarried embryos and foetuses tend to be socially invisible. While, as noted above, many people experience pregnancy loss as a bereavement similar to the death of other family members, there are few rituals available for them to commemorate and mourn the life, however short, that was lost. Other people may not perceive the dead embryo or foetus as a 'real person', viewing its loss rather as an unfortunate 'outcome' which will easily be replaced. They therefore fail to understand the grief felt by people who have experienced pregnancy loss or who have chosen to terminate a pregnancy because of a foetal anomaly (Davidson 2011, Frost *et al.* 2007, Keane 2009, Layne 2000, 2003a, 2003b, McCoyd 2007). Professional workers may tend to employ terminology which denies the humanity or baby-like features of unborn organisms. Thus, for example, they may use such terms as 'failed conception', 'products of conception', 'missed abortion' or 'reproductive wastage', or even 'dead embryo' or 'dead foetus' to refer to the entity that has perished. These technical, dehumanising terms can be extremely confronting for couples who are grieving the loss of their 'baby' (Jonas-Simpson and McMahon 2005).

Ways of commemorating and remembering an unborn death are often very important to people who have experienced miscarriage. This is

DOI: 10.1057/9781137310729

particularly the case when there is no 'body' to remember, take photographs of or mourn because the loss occurred early in pregnancy and it can be hard for people to define exactly what it is they are grieving for (Kevin 2011). People who have experienced pregnancy loss may save such memorabilia as positive result pregnancy test sticks, photographs of the woman's pregnant belly, ultrasound images, hand-prints and foot-prints made by the hospital on their behalf, birth certificates and photographs of the dead unborn as evidence to 'prove' that this 'baby' once existed, or use objects such as jewellery, plants and infants' clothing and toys to represent the lost child (Keane 2009, Layne 2000). In Japan an annual ceremonial memorial service, *mizuko kuyo*, for the aborted or miscarried unborn, the stillborn and dead children is performed, where people can pay their respects to the dead. As abortion rates are high in Japan, the majority of the 'water-children' who are mourned in this ceremony are aborted foetuses. Ritual prayers and offerings are made in memory of the 'water-children' and their parents offer apologies to them, allowing for a ritualised expression of guilt and atonement for the practice of abortion (Perrett 2000).

No such rituals are available in western countries, but since the advent of the internet, memorialisation websites, support groups for pregnancy loss, blogs and YouTube videos have allowed couples to discuss their feelings with others who have had similar experiences, and to present mementos of their dead embryos or foetuses online as a way of remembering them publicly and drawing attention to their existence. This can be a particularly important way for couples to give some materiality to their lost embryos or foetuses. The lost unborn are typically portrayed as 'angels' on these sites, a way for grieving parents to visibly represent the bodies of which there is little tangible evidence. The images of angels used of these sites tend to represent them as idealised chubby, healthy and typically white infants with wings, once again bestowing the embodiment and identity of the infant or child upon the embryo or foetus (Keane 2009). This concept of the 'lost' unborn as 'angels' was also noted by Layne (2003a) in her study of American women who had experienced miscarriage or stillbirth, even among those who did not explicitly refer to other Christian beliefs in their accounts. She argues that this terminology allows grieving women to give some kind of benevolent and beautiful shape to their lost child, especially for those whose pregnancy ends very early before they have a body they can view or hold.

DOI: 10.1057/9781137310729

YouTube videos featuring images of such unborn bodies regularly attract tens of thousands and even millions of views and vast numbers of expressions of sympathy in the comments section, in which they are again routinely refer to as 'angels'. Many of these memorialisation sites include photographs or video showing the parents holding and cuddling the dead foetus after its birth, wrapped or dressed as a living baby would be. Such sites therefore provide a view of the dead unborn body that previously was not available to most people. Where an infant was stillborn at or close to full-term it looks very similar in these images to a sleeping living newborn, although in some images the livid marks of decomposition begin to be evident. However those foetuses that were born at an earlier stage of gestation appear far from infant-like, despite attempts to render them as babies by wrapping them in blankets, giving them beanies to wear and positioning them near fluffy toys. Their skin is bright-red in colour, almost transparent, and their tiny bodies are thin, their minute limbs spindly.

If few rituals are available to commemorate and mourn the death of embryos or foetuses which were part of a pregnancy, even fewer exist to mark the death of *ex vivo* embryos that are surplus to IVF requirements or have been found to be non-viable, of 'low quality' or genetically abnormal via PGD and thus will never have the chance of being placed in a uterus. Some couples who have decided against donating their spare IVF embryos to other couples or for research purposes have chosen to give them 'funerals' or burial rites, by burying them in their garden, for example, thus creating a 'mourning object' in the case where there is no actual body, simply a blastocyst comprised of four microscopic cells (Ellison and Karpin 2011). Imperfect embryos, or those that are not high enough quality for implantation, are not accorded the same kind of procedures for couples to mark their loss and are merely disposed of by laboratory workers (Ellison and Karpin 2011).

Concluding comments

This chapter has shown the various ways in which unborn assemblages are configured across a range of social worlds, from the abortion clinic to the laboratory and IVF cryogenic storage facility to the YouTube memorial video. In all of these contexts the vitality and state of humanness accorded to the unborn have been integral to the ways in which they

DOI: 10.1057/9781137310729

are conceptualised and treated. Imputed value is also a central feature of the ways in which various kinds of unborn entities are treated and thought of, and this value may change as the entities move across social worlds. Wanted pregnancies and 'high quality' *ex vivo* embryos destined for uterine implantation attract value because of the precious child that will possibly result. Dead embryos or foetuses lost in miscarriage are also typically positioned as already children, and therefore as precious and angelic. Aborted unborn entities and surplus or defective IVF embryos may attract value for what they can offer medical research in cases where they would otherwise be summarily disposed of as 'clinical waste'. Value may be ascribed to surplus IVF embryos because of their positioning as potential children, offering hope to other infertile couples to have their own children, or because they are viewed as already children, perhaps the only children commissioning individuals will ever have. These entities may also hold value because of the time, money and effort involved in their creation. The next chapter will also address issues of value, this time in relation to the ways in which the unborn are configured as endangered in some way and the judgements made concerning what types of unborn entities should be accorded protection.

DOI: 10.1057/9781137310729

5
The Endangered Unborn

Abstract: *In the context of a growing emphasis on risk in contemporary developed societies, unborn entities sited in utero have increasingly become portrayed as 'at risk' from various dangers, most of which are viewed as conveyed to them via the actions of the maternal body in which they are developing. This chapter examines the intensification of discourses and practices that position the unborn as precious, vulnerable and endangered and pregnant women as ultimately responsible for their optimal health and development. The concepts of the 'reproductive citizen' and the 'foetal citizen' are introduced and it is argued that pregnant women's bodies have become increasingly subject to public surveillance. The chapter also identifies the ways in which pregnant women may be constituted as potentially abusing their unborn in legislation and how some embryos and foetuses are considered more worthy of protection than others.*

Key words: risk; pregnancy; foetal endangerment; reproductive citizen; foetal citizen; legislation

Lupton, Deborah. *The Social Worlds of the Unborn*. Basingstoke: Palgrave Macmillan, 2013.
DOI: 10.1057/9781137310729.

DOI: 10.1057/9781137310729

Risk and the reproductive citizen

Over history there have been constant anxieties and concerns expressed about what pregnant women eat, breathe in, drink and absorb and even with what they see, smell, wish and imagine and their emotional states, in relation to how these behaviours may affect the unborn (Dubow 2011, Hanson 2004, Kukla 2005). For example in the USA in the late nineteenth century, the prenatal movement put forward the argument that pregnant women should be careful of how they conducted themselves because of the effect their actions could have on their unborn. Both the mental and the physical states of the woman were viewed as potentially affecting the unborn entity growing within her, for either good or ill. Pregnant women were encouraged to avoid distressing or frightening sights, to think positive thoughts and read improving or inspiring literature and listen to high quality music, as these actions were viewed as having a direct influence on the personality and abilities of their unborn once it was born. During this period eugenicists were also arguing for the importance of ensuring that only people with high quality traits (typically glossed as wealthy or middle-class white people) should breed, and that the 'unfit' should be dissuaded from reproducing. Both perspectives focused on improving the quality of the unborn by intervening before conception or during pregnancy (Dubow 2011).

Until the middle of past century, although women's behaviours during pregnancy were considered to possibly shape their unborn's personality there was little understanding of the biological ways in which the maternal and the unborn body were interrelated. Until the 1960s the placenta was viewed in medicine as a kind of wall separating the foetus from the maternal body, screening out any toxins and sealing the unborn body from the maternal body. The pregnant woman's body and that of the unborn were seen therefore as largely physically independent of each other. Once the placenta became understood as permeable, however, the discourse of responsibility began to emerge, and pregnant women were represented as potentially posing a threat to their foetuses by behaviours which were seen to allow toxins or infecting agents to pass through the placental wall (Longhurst 2000a).

In the early decades of the twenty-first century, the beliefs that the unborn should be carefully monitored for their fitness to be born and that pregnant women should take care to engage in the appropriate behaviours to ensure the health and wellbeing of their unborn have

DOI: 10.1057/9781137310729

become ever more integral to understandings of unborn and pregnant embodiment. A discourse of risk has grown around the pregnant body, in which the subject 'at risk' is not the woman herself but the unborn entity growing inside her (Kukla 2010, Lupton 1999, 2011, 2012a, Lyerly *et al.* 2009, Weir 2006). As I observed in Chapter 1, children, and by extension the unborn, who are frequently positioned as proto-children or indeed as already children, have increasingly been portrayed as precious and vulnerable over the past half-century or so. Consonant with this representation is an intensified focus on the risks that are regarded as threatening children and the unborn (Beck and Beck-Gernsheim 1995, Kukla 2010, Lupton 1999, 2011, 2013a, 2013b, Lyerly *et al.* 2009, Ruhl 1999, Weir 2006).

The health and wellbeing of children of all ages is a central focus of the neoliberalist mode of politics that currently predominates in developed societies. Children represent future potentiality and the neoliberalist form of government is therefore concerned with fostering children's future capacities (Lee and Motzkau 2011). The unborn are positioned as even more defenceless and endangered than children, given their position within the potentially contaminating maternal body. Pregnant women have thus become a prime target for neoliberal governmental strategies directed not only at the 'care of the self', but even more importantly, the 'care of the (unborn) other': the valuable potential child. Indeed, for Duden, planet earth and the unborn became the primary symbols of endangered life in the 1990s: she refers to them as 'cult objects' (1993: 100). The concept of 'reproductive citizenship' (Salmon 2011: 167) denotes this emphasis on self-regulation in the interests of the health and wellbeing of the unborn. Pregnant women are viewed as conforming to the norms of good citizenship by taking up medical advice on how best to protect their unborn (future citizens) from harm.

The increasing dominance of the mode of medicine that has been termed 'surveillance medicine' (Armstrong 1995) is an important contributor to this intense focus on the behaviour and health of the pregnant woman and the wellbeing of her unborn. Under surveillance medicine, bodies are increasingly monitored, measured and compared against norms. In concert with the focus on self-responsibility for one's health status that is championed in neoliberalism (Lupton 1995), surveillance medicine renders the pregnant body a site of intense observation, requiring the close examination of medical professionals and their technologies but also the continual turning of the monitoring gaze by

DOI: 10.1057/9781137310729

pregnant women upon themselves. Both pregnant women and their foetuses have become biomedical subjects, their bodies defined, given meaning and regulated by the discourses of biomedicine. The prenatal screening and testing regimes that are offered to pregnant women and the regular prenatal appointments they attend with health professionals have become primary modes of surveillance of pregnant women and the unborn.

In discourses of pregnant embodiment and unborn risk, the notion that the mind should exert control over the body dominates. The pregnant woman is expected to exert her ethical responsibility by seeking out expert advice on how best to protect her unborn and subduing any bodily urges that might counteract this advice. As discussed in Chapter 3, the maternal body has increasingly been portrayed as permeable, unstable and emotionally volatile. Pregnant women are often represented as having little or no control over their bodies, and as captive to the influences of hormones. The notion of pregnant women's uncontrollable cravings for certain foodstuffs, or the representation of them as highly emotional and therefore irrational, also suggest that pregnant women are liable to lose control over what they allow in to their bodies, and therefore to what substances they expose their unborn.

In the discourses of reproductive citizenship the pregnant woman is expected to overcome her unruly body using her powers of rationality and moral reasoning. Pregnant women are therefore represented paradoxically as both responsible authorities and agents in the care of their own bodies and that of their unborn while simultaneously portrayed as liable to carelessly expose their unborn to danger (Ruhl 1999). They are expected by their health care providers to report any worries or anxieties they may have about their unborn's wellbeing as part of their regime of self-surveillance, but are also often positioned by these same health professionals as 'worrying too much' or becoming 'problem patients' if they express too many concerns (Bessett 2010). They must therefore negotiate demonstrating an appropriate level of concern about their own health and that of their unborn with avoiding over-anxiety or obsessiveness.

The notion of the vulnerable, precious, permeable unborn body as placed 'at risk' by the actions of the pregnant women in whose body it is developing has become more dominant with a growing medical literature investigating the effects on the unborn of maternal behaviours. This literature has been taken up in popular cultural representations of appropriate pregnant behaviour. As a study of articles on pregnancy

DOI: 10.1057/9781137310729

published in *The New York Times* and the American *People* magazine from 1996 to 2006 found, there was evidence of increasing reference to the 'risks', 'hazards' and 'perils' posed to the unborn during pregnancy. There was also increasing emphasis placed in these texts on the importance of preparing for conception months or even years ahead, with the possibility of adversely affecting one's children's health for the rest of their lives if this advice is not followed (Gentile 2011).

An increasing number of maternal behaviours have been identified as affecting the health and development of the unborn. In July 2011, an Australian weekend newspaper published a 'special report' on pregnancy in one of its supplements, including several articles on the topic. One of the articles in the supplement, bearing the headline 'Protecting precious cargo', concerned the practices in which pregnant women were advised to engage to ensure a healthy foetus (Prangell 2011). As it was noted in the subheadline, 'When you are pregnant, it's more important than ever to maintain a healthy mind and body.' In this article pregnant women were warned not only to avoid alcohol and tobacco, but also caffeine, fish with high mercury levels, soft cheeses, unpasteurised milk and other foods that may contain listeria bacteria, 'drastic diets', plastics containing BPA, radiation from x-rays, too much vitamin A and lying on their back in late pregnancy. They were further encouraged to drink enough water, get enough sleep, eat extra protein, take folic acid, Vitamin B and Omega-3 fatty acid supplements, to 'do your research' by reading pregnancy books to ensure that they are informed, participate in prenatal screening and testing to check for foetal abnormalities, exercise regularly, rest at work if it is physically demanding, engage in activities that are relaxing, be aware of potential mental health problems and that '[i]nfectious germs lurk everywhere, so keep this in mind when visiting doctors' surgeries or hospitals'.

Some recommended precautions and behaviours are directed at women even before they become pregnant, as part of the project of 'preparing' their bodies for pregnancy and ensuring that the resultant unborn entity will be optimally healthy and develop normally. In 2005 the Centers for Disease Control and Prevention and 35 partner organisations convened a national summit in the USA, and issued a set of recommendations concerning 'preconception care' for prospective parents (Frey *et al.* 2008). As part of 'preconception care' women planning pregnancy are advised to take appropriate vitamins, eat nutritious foods, exercise, avoid tobacco, alcohol and other drugs and certain medications, reduce their

DOI: 10.1057/9781137310729

body weight if deemed overweight or increase their weight if considered underweight, undergo tests for sexually transmissible diseases and other diseases such as hepatitis and toxoplasmosis and update their immunisations (Lanik 2012). More recently, a focus has also been placed on preparing men for prospective fatherhood in the preconception period, with men encouraged to take up similar lifestyle habits in the interests of ensuring that their sperm is optimal for conception and producing a healthy infant (Frey *et al.* 2008). These recommendations rely upon the construction of an imaginary identity: the 'preconceived embryo', 'a spectral figure whose form is imprinted on the imagined material body of the future child' (Karpin 2010: 136). They call upon both women and men, but particularly women, to engage in continual monitoring of their bodies and lifestyle habits for what may be a lengthy period of time (the 'preconception period') where no embryo yet exists and there is no guarantee that conception will occur.

Once conception has been successfully achieved and the presence of an unborn entity confirmed, vigilance over their bodies can by no means be relaxed for women but rather is intensified. The pregnant woman becomes positioned as 'the guardian of a public image' (Duden 1993: 54). By monitoring and regulating her own actions she is expected to create a shield of safety around her unborn by preventing any potentially polluting substances passing into the uterus. As was evident in the newspaper article I quoted, pregnant women are also expected to monitor their mental states, because the hormones associated with stress may affect their unborn adversely, and to avoid certain spaces that may contain 'invisible lurking germs' (particularly difficult when a pregnant woman is also expected to attend regular prenatal checks in apparently germ-infested doctors' surgeries or hospitals as part of her health regimen). In addition to these ascetic practices, pregnant women are also expected to 'read' their bodies so as to determine whether their unborn is healthy, monitoring foetal movements regularly as a sign of foetal life and reporting any symptoms such as abdominal pain or vaginal bleeding promptly to medical providers. All of these behaviours are part of how a 'good', 'responsible' and 'loving' mother-to-be is constructed (Bell *et al.* 2009, Lupton 1999, 2011, 2012a, Taylor 2000).

Part of this representation is the positioning of the pregnant woman as the threatening Other to the defenceless Self of the unborn. This configuration is particularly evident in such texts as Alexander Tsiaras's TED blog on his visualisations of the unborn (described in Chapter 2). In this

DOI: 10.1057/9781137310729

piece Tsiaras argues that the threat posed by pregnant women who fail to regulate their behaviour appropriately to their unborn has redefined 'our cherished image of the untainted and pristine embryo in the womb' (Tsiaras 2013: no page given). Tsiaras goes on to note that:

> I marveled at the symbiotic relationship that the developing child and mother shared. I saw only unbridled potential... the mother as a magnificent mobile heart/lung/immunology protector. Now I view pregnancy not as a perfectly loving mobile spa, but rather, as a fragile environment, one that must be kept healthy at all costs if a developing fetus is ever to be allowed to experience the delicate imperative of each of its genes. (2013: no page given)

As Tsiaras's words suggest, the 'awe' and 'wonder' that may be inspired by the unborn–maternal assemblage when it is conforming to the norms of medical expectations, the 'symbiotic' nature of this assemblage, may very quickly transform into fears about the threats that an unregulated maternal body may pose to the 'pristine' organism within. The pregnant body becomes a 'fragile environment' that must be kept as pure as possible in the interests of the unborn.

The emergence of the condition of 'foetal alcohol syndrome' in the early 1970s provides a case in point of how concepts of the pregnant and the unborn body have changed over the past generation. Foetal alcohol syndrome was the name used to describe a constellation of symptoms including pre- or post-natal growth retardation, central nervous system disorders and characteristic facial abnormalities that were linked to alcohol consumption of pregnant women. Although this diagnosis did not exist prior to the 1970s, it has become positioned as a major health problem, to the extent that pregnant women are now warned not to consume any alcohol during the course of their pregnancy because of possible adverse effects upon the unborn (Armstrong 1998, Bell *et al.* 2009). The intense focus on pregnant women's alcohol consumption serves to draw attention from other factors that influence unborn development and health, such as the nutritional status of the woman. Women of lower socioeconomic status who consume alcohol are far more likely to have an infant diagnosed with foetal alcohol syndrome than are more advantaged women who consumed the same amount of alcohol. This suggests that other factors related to socioeconomic disadvantage are influencing the development of the characteristics related to this syndrome (Armstrong 1998).

More recently, the emergence of what has been termed the 'obesity epidemic' has redirected attention to pregnant women's body weight. It

DOI: 10.1057/9781137310729

is now commonly asserted in the medical and public health literature that pregnant women who have been designated as 'overweight' or 'obese' place their unborn at risk by exposing them to the possibility of risks such as premature birth, complications during birth and stillbirth (because they may grow 'too big') and birth abnormalities, and following birth to a higher propensity for these children to develop obesity and diabetes (Gentile 2011, Keenan and Stapleton 2010, McNaughton 2011, Warin *et al.* 2012). Eating too little in the effort to remain thin during pregnancy is also considered inappropriate because it may lead to low birth weight or prematurity (Ralph *et al.* 2011). Thus, for example, when the pregnancy of Kate Middleton was announced, some commentators in the media asked whether she was 'too thin' to be pregnant and bear a healthy child (Rochman 2012). Pregnant women are therefore faced with treading a fine line between ensuring that their unborn will receive the kind of nutrition that is advised of them and maintaining what is considered to be a 'healthy' body weight. To eat too little or not enough of the 'right foods' would be considered just as risky to the unborn as to eat too much. For one's infant to be born 'too small' is equally as frowned upon as delivering 'too large' an infant.

The science of epigenetics has begun to place even more emphasis on the ways in which pregnant women's behaviour and even their mood may affect the unborn. According to a sub-branch of epigenetics, environmental epigenetics, the function or expression of the genetic code may be changed by environmental or lifestyle factors at the molecular biological level (Hedlund 2012). In environmental epigenetics it has been claimed that the physiological effects of stress upon the pregnant woman may change the genetic material of the unborn she carries within her, rendering it more likely to develop asthma or a mental illness later in life. These effects may even extend to the pre-conception period. It has been contended that both women's and men's gametes (their ova and sperm) may be altered by environmental conditions to which they are exposed, thus affecting future embryos. An even more specific area, nutritional epigenetics, focuses on the effects of diet on the developing unborn body in relation to changes at the molecular level in gene expression. The food consumed by the pregnant woman thus becomes an 'epigenetic factor' that may permanently affect the metabolic processes of the unborn organism, possibly leading to the child and even its descendants developing a propensity to such conditions as diabetes, cardiovascular disease, high blood pressure, obesity or metabolic syndrome (Landecker 2011).

DOI: 10.1057/9781137310729

Here is it not only maternal body weight that is viewed as problematic for the health of the unborn, but also the specific nutritional components of what she eats, regardless of her body weight.

In these discourses the preconception period and pregnancy become positioned as critical points at which intervention should take place. Epigenetics discourse, therefore, allows an even greater and more detailed focus on women's preconception and pregnancy behaviours. It again directs attention not at the pregnant woman's health or wellbeing but rather at those of her unborn. A pregnant woman's choices and those of a woman planning pregnancy have become linked to effects on her unborn that extend well into the future. The health and development of the unborn and even its descendants are positioned in these discourses as even more the product of the pregnant woman's lifestyle choices. Here again the role played by broader socioeconomic factors in shaping the health and development of the unborn is largely ignored.

The 'public pregnancy': women's experiences

Several writers have commented on the ways in which pregnant bodies become treated as public property once the distended abdomen signalling a foetus gestating within becomes obvious to onlookers (Kukla 2005, Longhurst 2000b, 2005, Nash 2013, Neiterman 2012). Kukla (2005) argues that the pregnant body is conceptualised as a public space in three senses. First, the inner space of the uterus has been externalised and subjected to scrutiny and women have come to understand what occurs in their uterus via third-party, mediated relationships such as those performed by doctors and prenatal testing technicians. Second, the pregnant body is a public domain in relation to how others treat it and attempt to monitor it. Third, there is a narrative space in which pregnancy unfolds, generated by medical and lay discourses on pregnancy, the pregnant body and the unborn, which provides guidelines for how pregnant women are expected to manage and monitor their bodies.

The concept of the 'baby bump', or the rounded pregnant abdomen containing and representing the unborn, has drawn attention to pregnant women's comportment of themselves both in public and private spaces (Nash 2013). Pregnant women are treated by other people differently, their bodies subject to far more scrutiny, comment and touching by others than any other adult body. During pregnancy the zone of corporeal

DOI: 10.1057/9781137310729

space around the body becomes thinner and may even disappear altogether. Other people, even strangers, feel as they have the right to touch the pregnant woman's 'bump' and to comment on it in ways they would never do in relation to other people's bodies (Longhurst 2000a, 2005, Nash 2013, Neiterman 2012). The protuberance of the pregnant abdomen, thus, becomes a public part of the woman's body, and the unborn body within is treated as if it is collective property, a 'potential citizen in which there is collective interest' (Longhurst 2000a: 58).

As the pregnant woman's body is treated as a vessel for the unborn to grow in, she is open to constant monitoring and critique from others on her behaviour, including the way she dresses, her body weight and the drugs, food and drink she chooses to consume (Longhurst 2000b, 2005, Lupton 2012a, Nash 2013, Neiterman 2012). This intense attention may become overwhelming for some women. English writer Rachel Cusk, in her memoir of pregnancy and early motherhood *A Life's Work: On Becoming a Mother* (2001), commented on her distress at her loss of privacy once her state of pregnancy was obvious to others because of the size of her abdomen. She describes her feelings that strangers had entered a room in her home and were 'riffling about' and that 'it is as if some spy is embedded within me, before whose scrutiny I am guilty and self-conscious'. This 'spy' she observes, is not the unborn presence in her body, but rather other people observing her behaviour and 'stating a claim' over her body (cited in Hanson 2004: 3).

Given the intense focus in lay and medical forums on the responsibility of pregnant women to discipline their bodies in the interests of their unborn, women who publicly engage in such activities as smoking or drinking alcohol are frequently subject to public castigation and disapproval. The obviously pregnant woman puffing on a cigarette or sipping a glass of wine or beer has become a deviant and often reviled figure because she is flouting societal expectations of the 'good/moral expectant mother' who is 'putting her baby first'. This negative portrayal was the focus of a BBC Three series first aired in 2011 entitled 'Misbehaving Mums to Be'. The series featured six pregnant women who had been chosen as case studies because they 'misbehave', or engage in behaviours deemed by medical authorities to pose a threat to their unborn: namely smoking, drinking alcohol, being overweight, not eating a nutritious diet and even working too many hours each day. The pregnant women who featured in the series were confronted by the midwives with warnings about how their choices would adversely affect their unborn and taken

DOI: 10.1057/9781137310729

on visits to places such as a hospital's Special Care infant unit to see newborn infants who were alleged to have been damaged by their mothers' behaviours while pregnant. As the BBC Three website describes the series, 'the midwives help these women transform their bad habits and reverse the dangers currently posed to their babies' lives' ('Misbehaving Mums to Be' 2012).

Empirical research from a range of developed countries, including the UK, the USA, Australia, Switzerland and Germany (Bessett 2010, Brown *et al.* 2010, Burton-Jeangros 2011, Erikson 2012, Gatrell 2011, Lupton 1999, 2008, 2011, Lyerly *et al.* 2009, Markens *et al.* 1997, Nash 2013, Root and Browner 2001) has demonstrated that many pregnant women are highly aware of and anxious about the possible risks faced to their unborn and attempt to change their behaviour in accordance with medical and public health advice. In Germany there is even a special term for the experience of 'pregnancy anxiety', a condition designated as requiring medical care (Erikson 2012). This research has found that regardless of the country in which they reside, it is common for pregnant women to express feelings of guilt if they feel that they have transgressed the behavioural norms expected of them and in so doing possibly placed their unborn at risk. Many pregnant women use a wide range of 'expert' resources to seek out information on how they should conduct themselves during pregnancy to ensure the optimum health and development of their unborn, including pregnancy handbooks and the multitude of websites offering advice for pregnant women as well as their medical caregivers (Gatrell 2011).

Pregnant women often articulate the discourse that the wellbeing of their unborn should be placed above their own needs or desires. The discourse of 'maternal self-sacrifice' or 'self-denial' is frequently used in pregnant women's accounts of how they discipline themselves in the interests of their unborn. Many observe that they gave up activities or food or beverages they enjoyed consuming pre-pregnancy as part of following expert advice (Gatrell 2011, Lupton 2011, Markens *et al.* 1997, Nash 2011, 2013). Pregnant women also frequently endure distressing symptoms such as debilitating nausea, heart-burn or back pain without complaint in the interests of their unborn (Bessett 2010, Nash 2013). Some women forfeit taking important medications required to protect their own health such as anti-depressants or asthma drugs, even when their doctors have advocated their use, for fear that these drugs will have unintended effects on their unborn (Lyerly *et al.* 2009).

DOI: 10.1057/9781137310729

In my own research with mothers of young children living in Sydney, I found that many of the women I interviewed noted that they had taken steps to protect their unborn while pregnant by taking vitamins, eating only healthy foods, avoiding foods with the risk of listeria, alcohol and smoking or smoke-filled places. Middle-class women were more likely to be vigilant about such actions, however, suggesting that some working-class women may not subscribe to the ideal of the risk-avoiding subject or are not able to because of socioeconomic disadvantage (Lupton 2008, 2011). In her interviews with Swiss pregnant women, Burton-Jeangros (2011) also found evidence of some women's resistance to the highly prescriptive discourses of reproductive citizenship. These women expressed their dismay and annoyance at the ways in which they were expected to give up practices they enjoyed, such as smoking or alcohol consumption. They were willing to challenge the intense risk-avoidance enjoined upon them by pointing to uncertainties in relation to some alleged risks and their effects on the unborn. Nonetheless, these women sometimes still expressed feelings of guilt for giving precedence to their own desires over the admonitions of medical advice.

Research has also found that pregnant women are often highly aware of the public censorious gaze levelled at them. A study of pregnant women in New Zealand, for example, found that many felt as if they should withdraw from public space because of self-consciousness about their bodies, physical discomfort, concerns about losing control over their bodies and the difficulty of conforming to expectations of how a 'proper' pregnant woman should comport herself (Longhurst 2000b). Pregnant women often report their behaviour as being 'policed' by others, including their partners and other family members or friends, as well as by strangers. One Australian woman reported in an interview that a taxi driver in whose vehicle she was travelling warned her about the dangers of drinking the coffee in the take-away cup she was clutching (Nash 2013). Some of the Swiss women in Burton-Jeangros' (2011) research observed how strangers had approached them in public to warn them about the risks to their unborn of behaviours such as using a mobile phone or smoking, and that their partners, friends or family members had advised them against drinking cola drinks or alcohol, consuming 'risky' foods or using cleaning chemicals. These women said that they were forced to confine such behaviours to the home, in secret, so as not to arouse the moral opprobrium of others.

DOI: 10.1057/9781137310729

Juggling the demands of the workplace, where the privations and needs of one's body are expected to be invisible, or at the very least, downplayed, and the demands of pregnancy ignored, can be difficult for women who are also attempting to conform to expert advice about adequate rest, diet and so on and who may be dealing with distressing conditions such as nausea and vomiting due to morning sickness (Gatrell 2011, Longhurst 2005, Nash 2013). In a sociocultural context in which fatness is viewed highly negatively, pregnant women must also negotiate their body weight, seeking to avoid putting on too much weight. This is both because fatness is viewed as unattractive or evidence of lack of physical fitness and, as noted earlier, because of the increasing focus on the relationship between maternal body weight and the future health of the unborn. As discussed in Chapter 3, pregnant women may find the increasing size of their bodies distressing because they feel as if they are 'too fat' and therefore unattractive, but they are also concerned to ensure that they eat nutritious foods for the sake of their unborn (Earle 2003, Gentile 2011, Lupton 2011, Markens *et al.* 1997, Nash 2011, 2013).

As a result of the focus on unborn health and wellbeing to the exclusion of pregnant women's interests and needs, and the discourse of responsibility and self-abnegation that permeates dominant cultural representations of pregnancy, women who suffer spontaneous pregnancy loss may define this as their fault (Frost *et al.* 2007, Kevin 2011). Indeed other pregnant women also often position those who have experienced miscarriage as responsible for this loss if they are judged to have not followed medical advice, for example, or rested enough or consumed an appropriate diet, and even if they are viewed as not having the right 'positive attitude' towards their unborn. Some women may also feel responsible for pregnancy loss because they left it 'too late' to conceive, given that medical advice warns that older women will suffer a higher rate of miscarriage (Kevin 2011, Root and Browner 2001).

The foetal citizen

Where once legal statues represented the unborn as part of a woman's body (Chapter 1), in several countries embryos and foetuses now have their own legal standing as citizens. In the USA, for example, recent changes in laws and regulations have demonstrated that the unborn are now recognised in that country as individual citizens. These

DOI: 10.1057/9781137310729

changes include the addition in 2002 of human embryos by the federal Department of Health and Human Services to the list of 'human subjects' whose welfare must be taken into account by the USA's Advisory Committee on Human Research Protection, the Department's redefinition that year of the term 'child' to begin at the moment of conception in revisions of the State Child Health Insurance Program and the passing by Congress in 2004 of the Unborn Victims of Violence Act, which rendered the death of a pregnant woman and her unborn entity in the execution of a federal crime punishable as two separate criminal violations (Dubow 2011). As I noted in Chapter 4, the recent 'personhood' initiatives in the USA have begun to argue even more insistently that the unborn should be considered fully moral and autonomous persons from the moment of conception.

The concept of the 'foetal citizen', bolstered by the emergence of foetal medicine and the subsequent portrayal of the unborn as patients, has been used to construct a legal entity that allowed foetal advocates to promote such causes as anti-abortion and the regulation of pregnant women's behaviour, including forcing medical care upon them against their wishes, in the interests of this citizen. In the 1970s, this new category of the 'foetal citizen' generated a movement devoted to fighting for 'foetal rights', conflicting with the emergent women's rights and related pro-abortion movement (Dubow 2011). From the 1960s onwards, as a consequence of the intensification of focus on women's responsibility for the health, wellbeing and optimal development of their unborn, pregnant women have been subjected to criminal prosecution for allegedly causing injury to their unborn by not seeking medically recommended prenatal care, continuing to take drugs or consume alcohol, having a car accident while intoxicated or refusing certain kinds of obstetric care. In Australia and the UK, women have been prosecuted for harm suffered to their unborn because of their negligent driving (Karpin 2010). In Canada, infants who have been born addicted to illicit drugs have been removed from their mothers by the courts. Foetuses have also been made wards of the state in that country because of judgements of instability in the pregnant woman or because she has refused a caesarean section that has been recommended by medical professionals (Ruhl 2002).

In the USA, legal interventions on behalf of the unborn are even more common. One study identified 413 cases from 1973 to 2005 in the USA where pregnant women were deprived of their liberty through arrests or forced medical interventions in the interests of their unborn (Paltrow

DOI: 10.1057/9781137310729

and Flavin 2013). Pregnant women have been ordered by US courts to undergo blood transfusions, cervical surgery or a caesarean section against their wishes, and have been detained in hospitals for enforced medical treatment or psychiatric care for drug addiction. In some cases the court has given temporary custody over the unborn child to the government, allowing it full authority to make all decisions concerning its welfare (Bordo 1993, Dubow 2011, Karpin 1992, Paltrow and Flavin 2013). Pregnant women have been prosecuted for child neglect, foetal abuse and manslaughter or homicide because they used drugs such as cocaine, amphetamines, heroin or even legal drugs such as alcohol while pregnant or failed to seek prenatal care. If they were convicted, some of these women have been forced to serve prison sentences (Dubow 2011, Paltrow and Flavin 2013, Ruddick 2007).

The vast majority of these women are black or Hispanic, or from the poor or working class. They are thus often dealing with social marginalisation and socioeconomic circumstances less than conducive to being able to conform to the high expectations of 'responsible motherhood' (Bordo 1993, Dubow 2011, Paltrow and Flavin 2013, Ruddick 2007, Ruhl 2002). In the 413 cases examined in the study conducted by Paltrow and Flavin (2013), in the majority of cases the women were deprived of their liberty even when no adverse effects on their unborn or infant of their behaviours could be demonstrated. In two out of three of these cases no health problems were found in the infants once they had been born, raising the question of how valid are the assumptions underpinning these actions.

Such cases often rely upon the discourse of child protection, and therefore constitute yet another way in which the unborn are positioned as already children. The unborn have even been rendered as independent 'parentless minors' in some legal discourses (Hartouni 1991: 28). These cases also represent such mothers as 'unnatural' in their failure to put their unborn's interests ahead of their own and protect their wellbeing, as 'good mothers' should (Dubow 2011, Ruddick 2007). It has been argued that these interventions may have the unintended negative consequence of deterring women from seeking prenatal care or revealing health conditions or drug use to their health care providers for fear of being arrested and punished, or even of encouraging women to seek abortions rather than risk being subjected to criminal penalties (Paltrow and Flavin 2013).

Many feminist critics have pointed out that in the wake of an intensification of focus in expert and popular forums on women's role in

DOI: 10.1057/9781137310729

protecting their unborn from risks, pregnant women and their unborn are represented as in opposition to each other, with the rights of the latter constantly privileged over those of the former (see, for example, Bell *et al.* 2009, Bordo 1993, Hartouni 1991, Karpin 1992, 2010, Kukla 2005, 2010, Lowe and Lee 2010, Lupton 2012a, McNaughton 2011, Petchesky 1987). In the discourses and practices of foetal endangerment, the unborn assemblage is represented as helplessly contained within the maternal assemblage, even as imprisoned in a body which is 'abusing' it (via the use of alcohol, tobacco or other drugs, for example). Medical and legal discourses that insist on the permeability of both the maternal and the unborn body tend to take an ambiguous and paradoxical position, simultaneously configuring these bodies as separate from each other but also as inextricably joined. While, for example, some laws position the unborn body as an individuated entity, thus allowing a woman to be sued following birth for injury experienced by the unborn, the very supposition that the actions of the mother caused damage to the unborn assumes that their bodies are connected and inseparable. Legal and medical discourses represent this connection, this inseparability, as a point of vulnerability for the unborn (Karpin 1992).

Feminist scholars have been highly critical of the ways in which these discourses of maternal responsibility position the unborn as more important than the pregnant woman. Some have contended that the mother-as-subject is erased by these discourses, predominantly represented as 'a mere life-support system for a fetus' (Bordo 1993: 77). This perspective on pregnant women is supported by the fact that there have been a number of cases in which brain-dead pregnant women have been kept alive for several weeks until the foetus was deemed mature enough to be delivered by caesarean section (Hartouni 1991). An instance of this occurred in 2012, when an American woman who was pregnant with twins and had suffered a brain aneurism which had destroyed her brain function was kept alive on a respirator for a month until her foetuses were delivered, following which the life support systems were turned off and she died (Hughes 2012). In such cases the woman's body is effectively dead flesh, kept alive by machines solely to act as an organic incubator for her foetus.

The ambiguous status of the 'two-in-one' unborn–maternal assemblage raises ethical issues for neoliberal governance, because the body of one citizen – the embryo or foetus – is located within the body of another – the pregnant woman (Weir 1996, 2006). The pregnant woman's

DOI: 10.1057/9781137310729

autonomy and freedom may be constrained by the presence of this other citizen body, but this is rarely acknowledged in discourses of foetal and reproductive citizenship. At its most extreme, as noted earlier, some pregnant women have been treated coercively and incarcerated in attempts to protect the unborn citizen. The voluntary nature of neoliberal government, therefore, is put aside in such cases for older-style coercive or punitive forms of rule. Furthermore, the rights and obligations discourse which underpins the individuation of the unborn from the maternal body positions the unborn as possessing none of the obligations and all of the rights of the typical liberal subject. The reverse is true of the pregnant woman. Indeed, in US legislation the unborn enjoy rights that are not accorded to adult liberal subjects, including the expectation that other people (specifically pregnant women) should subject themselves to heroic and often very invasive and risky medical treatment or surgery on their behalf (Ruhl 2002). In the USA pregnant women have fewer rights and are subjected to more extreme medical interventions in the interests of their unborn than are even cadavers, which cannot be used for medical purposes such as harvesting organs without the prior consent of the individual before death (Dubow 2011).

The intensity of focus upon pregnant women's actions and the assumed effects on the unborn bodies they carry serves to place great pressure upon women to conform to medical and public health advice. It is problematic, however, to assume that the health and development of the unborn is fully within the control of the pregnant woman. Although fate is no longer readily accepted as the reason for catastrophic outcomes in pregnancy, there are many aspects of unborn development and wellbeing that cannot be fully controlled by the individual. The discourses of reproductive and unborn citizenship, in privatising risk to the body of the pregnant woman, fail to recognise the sociocultural and political context in which pregnant women make choices about their behaviours – or indeed lack choices.

In championing the rights of the unborn citizen over those of the pregnant woman in whose body it is growing, attention is drawn away from the social and economic disadvantages faced by these women and the environmental health risks to the unborn. Women from socioeconomically disadvantaged backgrounds, for example, may struggle to conform to imperatives to control their behaviour in the interests of their unborn in a context in which they may experience physical, economic or emotional hardship or addiction to substances such as tobacco

DOI: 10.1057/9781137310729

and alcohol (Bell *et al.* 2009, Pickett *et al.* 2002, Salmon 2011). Potentially harmful contaminants such as environmental toxins are clearly beyond the individual's control (Kukla 2010, Markens *et al.* 1997, Ruhl 1999), yet the continual focus on the women's lifestyle 'choices' serves to obscure these contributors to maternal and unborn health.

Eugenics and the relative value of the unborn

It is important to point out that some unborn entities are represented as more precious than others in these discourses of foetal endangerment. Embryos or foetuses with diagnosed abnormalities or genetic disorders are considered less worthy of life in the medical arena: hence the emphasis on prenatal testing and screening technologies which seek to identify such abnormalities. The dominant rationale for encouraging pregnant women to undergo such tests and screens is to identify abnormalities or disease so that expecting parents can make an informed decision about whether or not to continue the pregnancy: in other words, to decide to dispose of the 'imperfect' embryo or foetus. The very existence of these technologies and their incorporation into routine prenatal care encourage pregnant women to undergo them and also imply that to do so is the appropriate way to approach pregnancy, part of endeavouring to produce the 'best' child one can.

As this suggests, the unborn entity that is in any way 'abnormal' is not as valued as those that are designated as conforming to normative standards. The unborn entity may be positioned as a 'suspect' under investigation until the results of prenatal screening or testing are confirmed (Ivry and Teman 2008). There is therefore a very clear eugenic principle underlying encouraging pregnant women to seek testing and screening of their unborn (Ettorre 2009). Such eugenic selection, weeding out the 'unfit' via abortion, is morally and legally sanctioned in many countries as a way of ensuring that infants are healthy and not disabled (Erikson 2007, Ettorre 2009, Gammeltoft 2007b, Gammeltoft and Nguyễn 2007, Gross 2010, McCoyd 2007). Taking this idea to its logical extreme, a number of utilitarian bioethicists have begun to champion what has been termed 'procreative beneficence', or the eugenic selection of embryos using PGD. One of the major proponents of procreative beneficence, Julian Savulescu (2001), contends that it is in fact a moral duty of prospective parents to use prenatal diagnostic technologies to

DOI: 10.1057/9781137310729

produce as high a quality child as they possibly can. He argues that such selection will ultimately benefit the child, its parents and society as a whole. According to Savulescu, PGD should not be limited to selecting embryos based on identifying the genes that predispose medical conditions, but also for genes that give a higher possibility of characteristics as high intelligence or propensity towards aggression, so that the resultant child will be able to have the 'best possible life'.

Some disability activists have constructed a trenchant challenge to the ways in which prenatal screening and testing technologies represent the unborn entity that has been diagnosed as potentially having or developing a disability as inferior and not worthy of life (Atkins 2008, Shakespeare 1999, Sharp and Earle 2002). Lay people have also expressed concerns that processes of evaluating and selecting the 'best quality' embryos or foetuses are unethical because of the discrimination it expresses towards people with disabilities (Lee 2000). Some individuals are particularly concerned that the use of selective abortion for sex selection positions foetuses who are the 'wrong' sex as unwanted and abnormal (Scully *et al.* 2006a, Scully *et al.* 2006b). As discussed in Chapter 4, many feminist writers are also critical of using prenatal testing technologies to select and terminate female embryos or foetuses. Some feminist critics are also concerned about the additional burden of reproductive responsibility that is placed upon women as part of the expectation that they will ensure a 'perfect' baby by engaging in medically recommended prenatal screening and testing procedures. They have suggested that women's decision to refuse these procedures may result in blame and responsibility being cast upon them in the case of an infant with disabilities or a medical condition being born (Atkins 2008, Ettorre 2009, Rapp 1990, Rothman 1994, Sharp and Earle 2002).

Foetuses that are developing in the 'wrong' type of maternal body may also be considered less worthy of life than others. In her analysis of representations of 'partial birth' (late term) abortions and 'crack babies' in the US, for example, Mason (2000) notes that the foetuses gestated within women who were addicted to crack were portrayed as less worthy of being 'saved' than were other foetuses. These 'crack babies', explicitly presumed to be black, were portrayed as tainted by their mothers' crack addition, as impure and polluted bodies that would impose a burden upon the state if they were born. Anti-abortion campaigners and politicians sought citizenship rights, privileges and protection for other foetuses – those involved in late-term abortions, who were routinely

DOI: 10.1057/9781137310729

presented as innocent victims of their mothers' whims to terminate them and as worthy of the privileges of citizenship. Such foetuses were represented as 'whole', 'normal' and free from disability or genetic conditions, implying that foetuses which were in some way disabled or with a genetic condition were not as worthy of their protection.

Here again, questions of value are integral to the configuration of the unborn assemblage. In a sociocultural context in which disability is highly stigmatised and viewed as a burden for the carers of children who are disabled, an 'imperfect' embryo or foetus is not considered worthy of life. Nor are the unborn that are developing within the bodies of women considered to be 'bad mothers' with 'uterine environments' that are not pristine. Such embryos and foetuses are positioned as expendable and worthless, far less considered 'foetal citizens' deserving of full human rights. Indeed it has been claimed that the imperfect unborn entity is positioned as the abject Other: that for which we feel revulsion because of its lack of normality. In such a context, the unborn's value is vastly reduced: instead of being considered precious, highly worthy of protection, it becomes viewed as lacking any value to the point that its life should be extinguished (McCabe and Holmes 2011).

Concluding comments

Discourses on reproductive and foetal citizens configure an unborn assemblage that is threatened by the maternal body in which it is developing. This configuration is influenced by the discourses and imagery emerging from the other social worlds in which the unborn move, which, as I have argued in previous chapters, are together producing a dominant unborn assemblage that is individuated, alienable from the maternal assemblage, a citizen in its own right. In other ways, however, the 'unborn at risk' discourse produces another kind of assemblage: that of the inextricable unborn–maternal assemblage, in which the actions and choices of the pregnant woman have a direct impact on the unborn contained within her. In discourses of reproductive and foetal citizenship, therefore, rather than the maternal body being erased from view, as it tends to be in visual imagery of the unborn, it is brought sharply into focus and relocated as the carrier of the unborn organism. Yet this assemblage tends to focus on the negative implications of this

DOI: 10.1057/9781137310729

interembodiment and interconnectedness of flesh, by representing it as a source of contagion or pollution of the pure, vulnerable unborn body by the contaminated pregnant body. The nurturing capacities of the maternal assemblage and the benefits bestowed upon the unborn by the women in whom they are developing are largely overlooked by these negative representations.

DOI: 10.1057/9781137310729

Final Thoughts

Abstract: *This brief conclusion brings together the main findings of the book, particularly in relation to the diversity and complexity of contemporary configurations of the unborn assemblage and the ways in which it interacts with the maternal assemblage. Alternative ways of thinking about the relationship between the unborn and maternal assemblages are suggested. This conclusion also discusses the debate in feminist theory between those who espouse the relational perspective on the unborn and the women who contribute to their creation and those who challenge it.*

Key words: feminist theory; pregnancy; foetuses; embryos; relational perspective; assemblages

Lupton, Deborah. *The Social Worlds of the Unborn.* Basingstoke: Palgrave Macmillan, 2013. DOI: 10.1057/9781137310729.

This book has identified various ways in which unborn entities are configured and the social worlds within which they reside and across which they move. A number of dominant trends in conceptualising, portraying and treating the unborn have emerged from the discussions presented herein. First, the ultimate authority of the pregnant woman in 'knowing' the unborn entity within her body evident in pre-modern times has been weakened by new medico–scientific knowledge, procedures and technologies. Second, the knowledge of the unborn entity bestowed by haptic sensations, which again were largely the preserve of the pregnant woman, have become superseded by the visual images offered by technologies that are able to look inside the uterus and even open it surgically to access the unborn body within. Third, the concept of the unborn/maternal body as a single entity has become progressively loosened, its connections weakened as the unborn have become increasingly positioned as individuated from the pregnant woman. Fourth, the pre-modern concept of the unborn as mysterious entities that lacked distinct personhood has changed to that of the unborn as already persons, and in many cases, as already infants. Fifth, unborn entities *in utero* have become portrayed as precious, pure, beautiful and vulnerable, subject to possible contamination from the maternal body within which they reside. And sixth, in the wake of IVF and its associated industries, as well as hESC and regenerative medicine, the unborn have been configured in some social worlds as commodities bearing commercial value, to be exchanged or sold, and in others as dehumanised therapeutic or research materials.

I have sought to demonstrate that whether an unborn entity is sited *in utero* or *ex utero*, a product of the laboratory or of human sexual activity, a wanted pregnancy or destined for termination, less than or greater than 14 days old, whole or disaggregated, gestated by a pregnancy surrogate or by the intended mother, whether it is deemed to be a baby or a mass of cells, alive or dead, normal or abnormal, high quality or low quality, viable or non-viable, human or therapeutic material, research material or clinical waste, female or male – all of these factors in various contexts influence the ways in which it is portrayed, treated and thought about. I have argued that different kinds of value are attributed to the contemporary unborn: as already or potential babies and members of the family, biomedical research resources, therapeutic agents offering hope to ill people or those with disabilities, commercial commodities, altruistic donations to help others bear their own children, the product

DOI: 10.1057/9781137310729

of women's labour in donating ova and acting as pregnancy surrogates, and individuals' emotional and financial investment in IVF. These meanings and values may change over the lifetime of an unborn entity. For example, an embryo that was originally created for the optimistic purpose of becoming 'a baby' for an infertile individual or couple may abruptly become potential clinical waste if it fails to develop normally, is found via PGD to be genetically defective or the 'wrong' sex or is surplus to the individual's or couple's fertility needs. However it may then again be reconfigured into precious research material and harbinger of therapeutic hope if it is used in hESC science, or back into a potential 'baby' if it is donated to another infertile couple for their IVF use.

I have further discussed the positioning of the unborn in some of the social worlds they inhabit as 'work objects'. For the pregnant women who are gestating them, the dominant cultural imperative to ensure that the unborn can be 'the best they can be' requires that they engage in ascetic and disciplinary practices of the self that may be interpreted as a kind of work. Commercial pregnancy surrogates and the parents who commission their services view gestation literally as a type of work, and the surrogates are paid accordingly. In the social worlds of the medical clinic and scientific laboratory, the unborn are configured as work objects in other ways as they are created, tested or screened, imaged and manipulated, their vitality brought into being, extinguished or transformed from one kind of matter to another.

Two of the dominant configurations of the unborn identified in this book position them as either innocently and passively threatened by the maternal body or else as apparently independent from this body. In these configurations the pregnant woman's body has tended to be either erased from the aestheticised image of the free-floating embryo or foetus or positioned as threatening Other to the pure, defenceless unborn body. In an attempt to bring back into focus the interests and subject position of pregnant women, various feminist writers have called for alternative representations. Some feminist scholars have sought to counter the disappearance of the maternal subject and the focus on foetal rights over those of the pregnant woman by presenting a relational account of unborn and maternal identities and embodiment (for example, Cohen 1999, 2000, Dickenson 2007, Holland 2001, Petchesky 1987, Ruhl 2002, Sherwin 1992, Warren 1989). These writers contend that unborn and maternal bodies should not be positioned as individuated but rather as always inextricably intertwined and interembodied. Some feminist theorists

DOI: 10.1057/9781137310729

have suggested highlighting images and discourses which emphasise the presence of the maternal body in harbouring and nurturing the unborn body, to reinstate the concept of 'two bodies in one' and the connectivity of these bodies (Hird 2007, Maher 2002, Shrage 2002). This alternative view positions the female reproducing body as unique, part of an ethic of generosity and intercorporeality in which the maternal body engages in a process of 'corporeal gifting' nourishment and vitality to the unborn body developing within her (Hird 2007: 5).

It has been further contended that the maternal body may be brought into the scene of gestation in a positive way by emphasising the vital role played by the placenta in nourishing the unborn and providing an important connection between unborn and the maternal bodies. From this perspective, the porosity and permeability of the unborn and maternal bodies become viewed not as evidence of inferiority or vulnerability, but rather of positive interdependence and mutual exchange (Hird 2007, Maher 2002). It is interesting to note that the dominant medical view of the pure and vulnerable unborn body as passively receiving substances from the maternal body in a one-way exchange has been challenged by recent research in the science of gestational cell transfer. This research has identified the movement via the placenta of foetal blood cells into the maternal blood stream and the subsequent mingling of these cells with maternal blood cells. This intermingling – entitled 'microchimerism' in the scientific literature – persists for many years following the pregnancy (perhaps permanently). This new way of thinking about and configuring the relationship between unborn and maternal bodies has implications for the politics of maternity and how pregnant women and the unborn are thought of as subjects. Like relational feminist philosophy, the discourse of microchimerism positions the unborn and maternal bodies as interrelated, interdependent and as both 'impure' in terms of their cellular and immunological integrity. Furthermore, the unborn body is viewed as active and agential, acting upon that of the mother rather than passively receiving whatever substances, good or bad, are transmitted from her body across the placenta (Hird 2007, Kelly 2012, Martin 2010).

Maternal–foetal microchimerism, therefore, could be described as an integral part of relationship of 'corporeal generosity', in which both unborn and maternal bodies are viewed as gifting substances to each other (Hird 2007). This is a very different way of conceptualising the interaction between the unborn and the pregnant woman. It avoids the suggestion of a hostile relationship which has dominated medical,

DOI: 10.1057/9781137310729

public health and legal portrayals of the foetal and maternal body to one in which interdependence, altruism and caring relations are paramount (Hird 2007, Kelly 2012, Martin 2010).

From the relational perspective the unborn body/self is inalienable from the maternal body/self: it cannot be understood to possess moral personhood or sentience distinct from the pregnant woman while it remains part of the maternal body and completely reliant upon this body. This position privileges the notion of the unified unborn–maternal assemblage. Proponents of this perspective are therefore highly critical of the notion that the embryos and foetuses produced by women should be commodified, or stripped of personhood and reduced to a marketable good. The practice of commercial surrogacy, for example, has incited several critiques concerning its ethical nature, particularly in relation to surrogates in poor countries employed by wealthy westerners to carry their children. Some feminists have argued that desperately poor women may be exploited by more wealthy people by being paid to be pregnancy surrogates, their bodies objectified by being rendered into little more than reproductive containers for the gestation of other people's children (Bailey 2011, Damelio and Sorensen 2008, Pande 2010).

This perspective, however, has in turn been critiqued by other feminists for being reductionist, by implying that relationality is biological rather than social (Conklin and Morgan 1996, Keane 2009, Layne 2000, Morgan 1996). These critics have contended that such a view on relationality supports the notion that unborn entities are not social and thus do not have social relationships with others. Bodies, in this view, are 'natural', while 'persons' are social. It is argued that the relational approach fails to acknowledge that, as this book has demonstrated, concepts of personhood develop earlier than the moment of physical birth via visualising technologies and the emotions and imaginings of expecting parents. The unborn, therefore, typically enter the social world as members in their own right well before they are physically born and separated from the maternal body. Some writers have asserted that the relational perspective also tends to obscure the grief and suffering felt by some women (and men) who experience miscarriage, because it does not acknowledge that such individuals view the embryo they lost as a 'person' with its own social identity separate from that of the woman who was gestating it (Keane 2009, Layne 2003a, 2003b, Morgan 1996).

Feminist theorists who do not agree with relational perspectives further contend that the view that embryos and foetuses are inalienable

DOI: 10.1057/9781137310729

parts of women's bodies limits women's reproductive freedom, including their right to terminate a pregnancy (Haimes *et al.* 2012, Mcleod and Baylis 2006). It does not recognise that some women may not see their ova, embryos or foetuses as inseparable parts of themselves. They argue, therefore, that women should be able to sell their ova, embryos and foetuses or dispose of them in any way they see fit. It has been contended further that placing an economic value on reproductive services such as ova donation and surrogate pregnancy serves to highlight their monetary worth and to acknowledge that these services are a form of work, a perspective that often receives little attention and acknowledgement in public debates on these topics (Berkhout 2008).

The research reviewed in this book suggests that both the relational perspective and the approach that calls for viewing the unborn as alienable from the maternal body/subject are reductive. As I have shown, pregnant women themselves often tend to articulate a shifting and ambivalent concept of their embodiment and selfhood in relation to their unborn. Their unborn are sometimes positioned as Other to one's Self and sometimes as an inextricable part of one's Self. In some contexts unborn entities may be considered alienable from the maternal subject, and therefore able to be traded or sold, and in other contexts as inalienable. These configurations may even change for the same woman during the term of her pregnancy. Such factors as the stage of progression of the pregnancy, the pregnancy symptoms the woman may experience, whether the pregnancy was wanted, accidental, part of a surrogacy arrangement or a product of rape, the types of foetal movements the woman feels, her use and interpretation of visualising technologies, her social, economic and cultural background, how she feels about her pregnant body and the attitudes and reactions of those around her can all serve to shape her concepts of her unborn in various ways.

It may be possible to reconcile both positions by acknowledging the dynamic nature of the ontology of unborn entities and its shifting meanings according to the context in which they are perceived and experienced. It is here that the concept of unborn and maternal assemblages works well to describe the dynamic ways in which the unborn are configured and reconfigured. This concept both recognises the relational and interembodied nature of this relationship and allows for constant shifts and changes that are dependent on the social worlds in which the unborn and maternal assemblages are located. Maternal and unborn assemblages come together and are joined in some contexts and in other

DOI: 10.1057/9781137310729

contexts come apart and are individuated from each other. Many pregnant women conceptualise their unborn as an integral part of their body
during the term of their pregnancy and find it difficult to think of the
unborn as an entity in its own right. Birth or pregnancy loss events are
the pivotal moment at which the unborn–maternal assemblage comes
apart in material, fleshly terms. Visualising technologies, as well as other
technologies and practices, in both the medical and lay worlds, that have
contributed to a dominant portrayal that depicts the unborn as separate
from the maternal body also work to disentangle these assemblages.
When unborn entities are destined never to enter a woman's body, as is
the case with surplus embryos created for IVF or those deemed defective
that may be then discarded or used for hESC research, they never enter
into a merged physical assemblage with the maternal body. However the
women who have commissioned the creation of these embryos may still
conceptualise them in some ways as 'part of me'.

The unborn, once understood solely as mysterious, hidden parts of
pregnant women's bodies, have become far more documented and complex entities over the past two centuries, and particularly in the past few
decades. During this period they have been invested with a range of often
conflicting values and their physical locations have diversified across
time and space. Only time will tell how unborn assemblages will be further transformed in response to new technologies, uses, understandings
and visualisations.

DOI: 10.1057/9781137310729

Glossary of Key Terms

Assemblage A term used in social and cultural theory to denote a bringing together of a diverse range of components to configure phenomena, including human bodies/selves, a process that is constantly shifting and changing depending on the context. This term incorporates recognition of the role played by non-human actors, including ideas, discourses, practices, spaces and material objects in configuring phenomena.

▶ **Blastocyst** The medical term used to refer to the conceptus from about day 5 to day 14 of development. The stage of development at which the *ex vivo* embryo is implanted into the uterus if used for assisted conception. The blastocyst is also sometimes referred to as the 'pre-14 day' embryo or 'pre-embryo'.

Cryopreservation The process used to preserve sperm, ova and *ex vivo* embryos using freezing technologies.

Embryo The word used for the organism created by fertilisation of an ovum with a sperm until the end of the eighth week of development post-fertilisation (ten weeks gestational age). The embryo is sometimes more specifically termed the 'zygote' from fertilisation until about day four of development, when the cells begin to divide, and the 'blastocyst' from about day five to day 14 of development.

***Ex vivo* embryo** An embryo created and located outside the female human body. A term used to refer to embryos created in the laboratory via IVF or cloning techniques.

Foetus The term used for the organism post-embryo stage, from the ninth week of development (eleventh week of gestation) until birth.

DOI: 10.1057/9781137310729

Gestational age The method used in medicine to date the stage of development of embryos and foetuses. The age in days or weeks of the conceptus from the date of the pregnant woman's last menstrual period, calculated as occurring two weeks (14 days) before conception. This age is therefore two weeks longer than the actual age of development of the embryo/foetus post fertilisation.

Human embryonic stem cells (hESCs) Cells taken from human embryos at the blastocyst stage of development for use in regenerative medical research and therapeutic treatment.

Induced pluripotent stem cell (iPSC) An adult stem cell that has been artificially altered into a pluripotent stem cell, currently under investigation for the possibility that it may replace hESCs.

In vitro **fertilisation (IVF)** The process by which human ova are fertilised by human sperm in the laboratory to create embryos (also referred to in popular parlance as 'test tube babies' and in medicine as *ex vivo* embryos).The highest quality blastocysts that result are selected for implantation into the uterus to establish a pregnancy. IVF is used to assist infertile individuals achieve pregnancy and also for the purpose of performing preimplantation genetic diagnosis (PGD) and in ova donation and surrogate pregnancy arrangements.

In vivo **embryo** An embryo conceived and located within the female body.

Non-embryo The entity that is made from techniques such as cloning or induced parthenogenesis and therefore does not involve the uniting of a male and female gamete or a unique genetic blueprint.

Pre-embryo A scientific technical term referring to the product of conception before the 14th day of development post-fertilisation.

Preimplantation genetic diagnosis (PGD) A genetic diagnostic technique whereby embryos created *in vitro* are tested at the blastocyst stage to determine such factors as their sex, whether they have any genetic disorders or chromosomal abnormalities or to check their suitability to provide organs or cells to treat a sibling with a serious medical condition. The results of PGD are used to decide which embryos are the most optimal to be implanted in the uterus of the commissioning woman.

Regenerative medicine A field of medicine in which scientific techniques are used therapeutically to replace or regenerate damaged or malfunctioning human cells, tissues or organs. Some techniques use human embryonic stem cells (hESCs) to attempt this.

DOI: 10.1057/9781137310729

Surrogate pregnancy This occurs when a woman is commissioned by an individual or couple either to be artificially inseminated, conceiving with her own ova, or (more commonly) to have an embryo implanted into her uterus and then to gestate the resultant pregnancy until birth, following which she relinquishes the infant to the commissioning individual or couple. In altruistic surrogacy the surrogate is not paid except for fees to cover her expenses, while in commercial surrogacy she does receive additional fees for her services.

DOI: 10.1057/9781137310729

Web Resources

'Making Visible Embryos' (http://www.hps.cam.ac.uk/visibleembryos/index.html): an online exhibition developed by the University of Cambridge with support from the Wellcome Trust about the history of images of human embryos.

'Sociology of the Unborn' Pinterest board (http://pinterest.com/dalupton/sociology-of-the-unborn/): images I have collected of embryos and foetuses in medical and popular culture.

▶ 'The Embryo Project Encyclopedia' (http://embryo.asu.edu/index.php): a resource directed at understanding and identifying the agents of change shaping embryo research and its contexts, including encyclopedia entries, scholarly essays, images, timeline visualisations and videos.

'The Ultrasound as Cultural Artefact' Pinterest board (http://pinterest.com/dalupton/the-ultrasound-as-cultural-artefact/): images I have collected on obstetric ultrasounds in medical and popular culture.

'The Unborn Human' Scoop.it collection (http://www.scoop.it/t/the-unborn-human): digital news articles and blog posts I have collected that discuss issues concerning embryos and foetuses.

DOI: 10.1057/9781137310729

Bibliography

Abboud, L. and Liamputtong, P. (2002) Pregnancy loss. *Social Work in Health Care*, 36 (3), 37–62.

Adams, A. (1993) Out of the womb: the future of the uterine metaphor. *Feminist Studies*, 19 (2), 269–89.

Adams, S. (2008) Flayed babies' bodies included in new Body World exhibition. *The Telegraph*, Accessed 10 April 2012. Available from http://www.telegraph.co.uk/news/3249044/Flayed-babies-bodies-included-in-new-Body-World-exhibition.html.

▶ Alhusen, J. (2008) A literature update on maternal–fetal attachment. *Journal of Obstetric, Gynecologic, & Neonatal Nursing*, 37 (3), 315–28.

Anonymous (2012) *Sad Abortion Video* Accessed 12 September 2012. Available from http://www.youtube.com/watch?v=5x U2jqDD60A&feature=player_detailpage.

Armstrong, D. (1995) The rise of surveillance medicine. *Sociology of Health & Illness*, 17, 393–404.

Armstrong, E.M. (1998) Diagnosing moral disorder: the discovery and evolution of fetal alcohol syndrome. *Social Science & Medicine*, 47 (12), 2025–42.

Atkins, C. (2008) The choice of two mothers: disability, gender, sexuality, and prenatal testing. *Cultural Studies – Critical Methodologies*, 8 (1), 106–29.

Avalos, L. (1999) Hindsight and the abortion experience: what abortion means to women years later. *Gender Issues*, 17 (2), 35–57.

Bailey, A. (2011) Reconceiving surrogacy: toward a reproductive justice account of Indian surrogacy. *Hypatia*, 26 (4), 715–41.

DOI: 10.1057/9781137310729

Baylis, F. (2000) Our cells/ourselves: creating human embryos for stem cell research. *Women's Health Issues*, 10 (3), 140–45.

Beck, U. and Beck-Gernsheim, E. (1995) *The Normal Chaos of Love*. Cambridge: Polity.

Bell, K., McNaughton, D. and Salmon, A. (2009) Medicine, morality and mothering: public health discourses on foetal alcohol exposure, smoking around children and childhood overnutrition. *Critical Public Health*, 19 (2), 155–70.

Berkhout, S. (2008) Buns in the oven: objectification, surrogacy, and women's autonomy. *Social Theory and Practice*, 34 (1), 95–117.

Bessett, D. (2010) Negotiating normalization: the perils of producing pregnancy symptoms in prenatal care. *Social Science & Medicine*, 71 (2), 370–77.

Betterton, R. (2002) Prima gravida. *Feminist Theory*, 3 (3), 255–70.

Beynon-Jones, S.M. (2012) Timing is everything: the demarcation of 'later' abortions in Scotland. *Social Studies of Science*, 42 (1), 53–74.

Bindley, K. (2013) Ultrasound parties: sonograms with friends grow in popularity, become latest pregnancy trend. *Huff Post Parents*, Accessed 2 January 2013. Available from http://www.huffingtonpost.com/2013/01/02/ultrasound-parties-sonogram-viewings-pregnancy-trend_n_2397271.html.

Birenbaum-Carmeli, D., Yoram, S.C. and Rina, C. (2000) Our first 'IVF baby': Israel and Canada's press coverage of procreative technology. *International Journal of Sociology and Social Policy*, 20 (7), 1–38.

BodyWorlds (2013) Accessed 27 October 2012. Available from www.bodyworlds.com/.

Bordo, S. (1993) *Unbearable Weight: Feminism, Western Culture, and the Body*. Berkeley, CA: University of California Press.

Boucher, J. (2004) Ultrasound: a window to the womb? Obstetric ultrasound and the abortion rights debate. *Journal of Medical Humanities*, 25 (1), 7–19.

Brookman-Amissah, E. (2012) Saving women's lives in Africa through access to comprehensive abortion care. *The European Journal of Contraception & Reproductive Health Care*, 17 (4), 241–44.

Brown, J.E., Broom, D.H., Nicholson, J.M. and Bittman, M. (2010) Do working mothers raise couch potato kids? Maternal employment and children's lifestyle behaviours and weight in early childhood. *Social Science & Medicine*, 70 (11), 1816–24.

DOI: 10.1057/9781137310729

Burton-Jeangros, C. (2011) Surveillance of risks in everyday life: the agency of pregnant women and its limitations. *Social Theory & Health*, 9 (4), 419–36.

Carmosy, C. (2012) Concern for our vulnerable prenatal and neonatal children: a brief reply to Giubilini and Minerva. *Journal of Medical Ethics*, Accessed 19 November 2012. Available from http://jme.bmj.com/content/early/2012/03/01/medethics-2011-100411/suppl/DC3.

Carter, S. (2010) Beyond control: body and self in women's childbearing narratives. *Sociology of Health & Illness*, 32 (7), 993–1009.

Casper, M. (1994a) At the margins of humanity: fetal positions in science and medicine. *Science, Technology & Human Values*, 19 (3), 307–23.

Casper, M. (1994b) Freframing and grounding nonhuman agency: what makes a fetus an agent? *American Behavioral Scientist*, 37 (6), 839–56.

Casper, M. (1998) *The Making of the Unborn Patient: A Social Anatomy of Fetal Surgery*. New Jersey: Routledge University Press.

Casper, M. and Morrison, D. (2010) Medical sociology and technology: critical engagements. *Journal of Health and Social Behavior*, 51 (S), S120–S132.

Catechism of the Catholic Church: Part Three – Life in Christ (2013) Accessed 19 January 2013. Available from http://www.vatican.va/archive/ccc_css/archive/catechism/p3s2c2a5.htm.

Cesarino, L. and Luna, N. (2011) The embryo research debate in Brazil: from the National Congress to the Federal Supreme Court. *Social Studies of Science*, 41 (2), 227–50.

Chervenak, F. and McCullough, L. (2011) Ethics in obstetric altrasound: the past 25 years in perspective. *Donald School Journal of Ultrasound in Obstetrics and Gynecology*, 5 (2/3), 79–84.

Christoffersen-Deb, A. (2012) Viability: a cultural calculus of personhood at the beginnings of life. *Medical Anthropology Quarterly*, 26 (4), 575–94.

Cohen, C. (1999) Selling bits and pieces of humans to make babies: 'the gift of the magi revisited'. *Journal of Medicine and Philosophy*, 24 (3), 288–306.

Cohen, C. (2000) Use of 'excess' human embryos for stem cell research: protecting women's rights and health. *Women's Health Issues*, 10 (3), 121–26.

Cohen, C.B. (1996) 'Give me children or I shall die!' New reproductive technologies and harm to children. *The Hastings Center Report*, 26 (2), 19–27.

Collins, L.R. and Crockin, S.L. (2012) Fighting 'personhood' initiatives in the United States. *Reproductive Biomedicine Online*, (7). Accessed 5 October 2012. Available from http://linkinghub.elsevier.com/retrieve /pii/S147264831200209X?showall=true.

Conklin, B. and Morgan, L. (1996) Babies, bodies, and the production of personhood in North America and a native Amazonian society. *Ethos*, 24 (4), 657–94.

Corea, G. (1985) *The Mother Machine: Reproductive Technologies from Artificial Insemination to Artifical Wombs*. New York: Harper and Row.

Crow, B.L. (2010) Bare-sticks and rebellion: the drivers and implications of China's reemerging sex imbalance. *Technology in Society*, 32 (2), 72–80.

Cusk, R. (2001) *A Life's Work: On Becoming a Mother*. London: Fourth Estate.

Damelio, J. and Sorensen, K. (2008) Enhancing autonomy in paid surrogacy. *Bioethics*, 22 (5), 269–77.

Davidson, D. (2011) Reflections on doing research grounded in my experience of perinatal loss: from auto/biography to autoethnography. *Sociological Research Online*, (1). Accessed 3 April 2012. Available from http://www.socresonline.org.uk/16/1/6.html.

de Lacey, S. (2007) Decisions for the fate of frozen embryos: fresh insights into patients' thinking and their rationales for donating or discarding embryos. *Human Reproduction*, 22 (6), 1751–58.

Dickenson, D. (2007) *Property in the Body: Feminist Perspectives*. Cambridge: Cambridge University Press.

Diepper, S. (2012) The legal framework of abortions in Germany. Accessed 13 January 2013. Available from http://www.aicgs.org/issue /the-legal-framework-of-abortions-in-germany/.

Douglas, M. (1969) *Purity and Danger: An Analysis of Concepts of Pollution and Taboo*. London: Routledge & Kegan Paul.

Draper, J. (2002) 'It was a real good show': the ultrasound scan, fathers and the power of visual knowledge. *Sociology of Health & Illness*, 24 (6), 771–95.

Draper, J. (2003) Blurring, moving and broken boundaries: men's encounters with the pregnant body. *Sociology of Health & Illness*, 25 (7), 743–67.

Dubow, S. (2011) *Ourselves Unborn: A History of the Fetus in Modern America*. New York: Oxford University Press.

DOI: 10.1057/9781137310729

Duden, B. (1993) *Disembodying Women: Perspectives on Pregnancy and the Unborn.* Translated by L. Hoinacki. Cambridge, MA: Harvard University Press.

Earle, S. (2003) 'Bumps and boobs': fatness and women's experiences of pregnancy. *Women's Studies International Forum,* 26 (3), 245–52.

Early View Ultrasound Centre (2013) Accessed 5 January 2013. Available from http://www.earlyview.com.au/.

Ehrenreich, B. and English, D. (1974) *Complaints and Disorders: The Sexual Politics of Sickness.* London: Compendium.

Ehrich, K., Williams, C. and Farsides, B. (2008) The embryo as moral work object: PGD/IVF staff views and experiences. *Sociology of Health & Illness,* 30 (5), 772–87.

Ehrich, K., Williams, C. and Farsides, B. (2010) Fresh or frozen? Classifying 'spare' embryos for donation to human embryonic stem cell research. *Social Science & Medicine,* 71 (12), 2204–11.

Ehrich, K., Williams, C., Farsides, B., Sandall, J. and Scott, R. (2007) Choosing embryos: ethical complexity and relational autonomy in staff accounts of PGD. *Sociology of Health & Illness,* 29 (7), 1091–106.

Ehrich, K., Williams, C., Farsides, B. and Scott, R. (2012) Embryo futures and stem cell research: the management of informed uncertainty. *Sociology of Health & Illness,* 34 (1), 1114–29.

Ehrich, K., Williams, C., Scott, R., Sandall, J. and Farsides, B. (2006) Social welfare, genetic welfare? Boundary-work in the IVF/PGD clinic. *Social Science & Medicine,* 63 (5), 1213–24.

Ellison, D. and Karpin, I. (2011) Embryo disposition and the new death scene. *Cultural Studies Review,* 17 (1), 81–100.

Embryo Bank (2012) Accessed 9 December 2012. Available from http://www.ivf-worldwide.com/Education/embryo-bank.html.

Embryo Princess (2012) Accessed 4 April 2012. Available from http://adventuretime.wikia.com/wiki/Embryo_Princess.

Erikson, S.L. (2007) Fetal views: histories and habits of looking at the fetus in Germany. *Journal of Medical Humanities,* 28 (4), 187–212.

Erikson, S.L. (2012) Social embodiments: prenatal risk in postsocialist Germany. *Anthropologica,* 54 (1), 83–94.

Ettorre, E. (2009) Prenatal genetic technologies and the social control of pregnant women: a review of the key issues. *Marriage & Family Review,* 45 (5), 448–68.

Featherstone, L. (2008) Becoming a baby? The foetus in late nineteenth-century Australia. *Australian Feminist Studies,* 23 (58), 451–65.

DOI: 10.1057/9781137310729

Franklin, S. (1997) *Embodied Progress: A Cultural Account of Assisted Conception*. London: Routledge.

Franklin, S. (2006a) The cyborg embryo: our path to transbiology. *Theory, Culture & Society*, 23 (7–8), 167–87.

Franklin, S. (2006b) Embryonic economies: the double reproductive value of stem cells. *BioSocieties*, 1 (1), 71–90.

Frey, K.A., Navarro, S.M., Kotelchuck, M. and Lu, M.C. (2008) The clinical content of preconception care: preconception care for men. *American Journal of Obstetrics and Gynecology*, 199 (6 Suppl 2), S389–95.

Frost, J., Bradley, H., Levitas, R., Smith, L. and Garcia, J. (2007) The loss of possibility: scientisation of death and the special case of early miscarriage. *Sociology of Health & Illness*, 29 (7), 1003–22.

Gammeltoft, T. (2007a) Prenatal diagnosis in postwar Vietnam: power, subjectivity, and citizenship. *American Anthropologist*, 109 (1), 153–63.

Gammeltoft, T. (2007b) Sonography and sociality: obstetrical ultrasound imaging in urban Vietnam. *Medical Anthropology Quarterly*, 21 (2), 133–53.

Gammeltoft, T. and Nguyễn, H.T.T. (2007) Fetal conditions and fatal decisions: ethical dilemmas in ultrasound screening in Vietnam. *Social Science & Medicine*, 64 (11), 2248–59.

Ganchoff, C. (2004) Regenerating movements: embryonic stem cells and the politics of potentiality. *Sociology of Health & Illness*, 26 (6), 757–74.

Gatrell, C. (2011) Putting pregnancy in its place: conceiving pregnancy as carework in the workplace. *Health & Place*, 17 (2), 395–402.

Gentile, K. (2011) What about the baby? The new cult of domesticity and media images of pregnancy. *Studies in Gender and Sexuality*, 12 (1), 38–58.

Georges, E. (1996) Fetal ultrasound imaging and the production of authoritative knowledge in Greece. *Medical Anthropology Quarterly*, 10 (2), 157–75.

Gerber, E. (2002) Deconstructing pregnancy: RU486, seeing 'eggs', and the ambiguity of very early conceptions. *Medical Anthropology Quarterly*, 16 (1), 92–108.

Ginsburg, F. (1990) The 'word-made' flesh: the disembodiment of gender in the abortion debate. In F. Ginsburg and A.L. Tsing (eds) *Uncertain Terms: Negotiating Gender in American Culture*. Boston, MA: Beacon Press, 59–73.

DOI: 10.1057/9781137310729

Giubilini, A. and Minerva, F. (2012) After-birth abortion: why should the baby live? *Journal of Medical Ethics*, Accessed 12 October 2012. Available from http://jme.bmj.com/content/early/2012/03/01/medethics-2011-100411.abstract.

God's Little Ones (2012) Accessed 31 March 2012. Available from http://www.godslittleones.com/.

Goslinga-Roy, G. (2000) Body boundaries, fiction of the female self: an ethnographic perspective on power, feminism, and the reproductive technologies. *Feminist Studies*, 26 (1), 113–40.

Gottlieb, A. (2000) Where have all the babies gone? Toward an anthropology of infants (and their caretakers). *Anthropological Quarterly*, 73 (3), 121–32.

Greasley, K. (2012) Abortion and regret. *Journal of Medical Ethics*, 38 (12), 705–11.

Gross, M. (2002) Abortion and neonaticide: ethics, practice and policy in four nations. *Bioethics*, 16 (3), 202–30.

Gross, S. (2010) 'The alien baby': risk, blame and prenatal indeterminacy. *Health, Risk & Society*, 12 (1), 21–31.

Grosz, E. (1994) *Volatile Bodies: Toward a Corporeal Feminism*. Sydney: Allen & Unwin.

Gupta, J.A. (2011) Exploring appropriation of surplus' ova and embryos in Indian IVF clinics. *New Genetics and Society*, 30 (2), 167–80.

Hadley, J. (1994) God's bullies: attacks on abortion. *Feminist Review*, (48), 94–113.

Haimes, E., Porz, R., Scully, J. and Rehmann-Sutter, C. (2008) 'So, what is an embryo?': a comparative study of the views of those asked to donate embryos for hESC research in the UK and Switzerland. *New Genetics & Society*, 27 (2), 113–26.

Haimes, E., Taylor, K. and Turkmendag, I. (2012) Eggs, ethics and exploitation? Investigating women's experiences of an egg sharing scheme. *Sociology of Health & Illness*, 34 (8), 1199–214.

Han, S. (2008) Seeing like a family: fetal ultrasound images and imaginings of kin. In J.M. Law and V. Sasson (eds) *Imagining the Fetus: the Unborn in Myth, Religion, and Culture*. Oxford: Oxford University Press, 275–90.

Hanson, C. (2004) *A Cultural History of Pregnancy: Pregnancy, Medicine and Culture, 1750–2000*. Houndmills: Palgrave Macmillan.

Haraway, D. (1991) *Simians, Cyborgs and Women: The Reinvention of Nature*. New York: Routledge.

DOI: 10.1057/9781137310729

Harpel, T. and Hertzog, J. (2010) 'I thought my heart would burst':
the role of ultrasound technology on expectant grandmotherhood.
Journal of Family Issues, 31 (2), 257–74.

Harris, G., Connor, L., Bisits, A. and Higginbotham, N. (2004) 'Seeing
the baby': pleasures and dilemmas of ultrasound technologies for
primiparous Australian women. *Medical Anthropology Quarterly*, 18
(1), 23–47.

Hartouni, V. (1991) Containing women: reproductive discourse in the
1980s. In C. Penley and A. Ross (eds) *Technoculture*. Minneapolis,
MN: University of Minnesota Press, 27–56.

Hartouni, V. (1992) Fetal exposures: abortion politics and the optics of
allusion. *Camera Obscura*, 10 (2), 130–49.

Harvey, O. (2005) Regulating stem-sell research and human cloning
in an Australian context: an exercise in protecting the status of the
human subject. *New Genetics and Society*, 24 (2), 125–35.

Hashiloni-Dolev, Y. and Shkedi, S. (2007) On new reproductive
technologies and family ethics: pre-implantation genetic diagnosis
for sibling donor in Israel and Germany. *Social Science & Medicine*, 65
(10), 2081–92.

Hashiloni-Dolev, Y. and Weiner, N. (2008) New reproductive
technologies, genetic counselling and the status of the fetus: views
from Germany and Israel. *Sociology of Health & Illness*, 30 (7), 1055–69.

Hays, S. (1996) *The Cultural Contradictions of Motherhood*. New Haven,
CT: Yale University Press.

Hazell, K. (2012) Five million IVF babies born since 1978. *HuffPost
Lifestyle United Kingdom*, Accessed 21 November 2012. Available from
http://www.huffingtonpost.co.uk/2012/07/02/health-ivf-five-million-
babies-1978_n_1642231.html.

Hedlund, M. (2012) Epigenetic responsibility. *Medicine Studies*, 3 (3),
171–83.

Hessini, L. (2007) Abortion and Islam: policies and practice in the Middle
East and North Africa. *Reproductive Health Matters*, 15 (29), 75–84.

Hird, M. (2007) The corporeal generosity of maternity. *Body & Society*,
13 (1), 1–20.

Hochschild, A. (2011) Emotional life on the market frontier. *Annual
Review of Sociology*, 37, 21–33.

Hoeyer, K., Nexoe, S., Hartlev, M. and Koch, L. (2009) Embryonic
entitlements: stem cell patenting and the co-production of
commodities and personhood. *Body & Society*, 15 (1), 1–24.

DOI: 10.1057/9781137310729

Hoggart, L. (2012) 'I'm pregnant... what am I going to do?' An examination of value judgements and moral frameworks in teenage pregnancy decision making. *Health, Risk & Society*, 14 (6), 533–49.

Holland, S. (2001) Contested commodities at both ends of life: buying and selling gametes, embryos, and body tissues. *Kennedy Institute of Ethics Journal*, 11 (3), 263–84.

Hopkins, N., Zeedyk, S. and Raitt, F. (2005) Visualising abortion: emotion discourse and fetal imagery in a contemporary abortion debate. *Social Science & Medicine*, 61 (2), 393–403.

Hubbard, R. (1984) Personal courage is not enough: some hazards of childbearing in the 1980s. In R. Arditti, R. Klein and S. Minden (eds) *Test-tube Women: What Future for Motherhood?* London: Pandora, 331–55.

Hughes, M. (2012) Brain-dead mother has twins after being kept alive. *The Daily Telegraph*, 24 April, 17.

Ikemoto, L. (2009) Eggs as capital: human egg procurement in the fertility industry and the stem cell research enterprise. *Signs*, 34 (4), 763–81.

Ivry, T. and Teman, E. (2008) Expectant Israeli fathers and the medicalized pregnancy: ambivalent compliance and critical pragmatism. *Culture, Medicine, and Psychiatry*, 32 (3), 358–85.

Jackson, E. (2006) Fraudulent stem cell research and respect for the embryo. *BioSocieties*, 1 (3), 349–56.

James, W.R. (2000) Placing the unborn: on the social recognition of new life. *Anthropology & Medicine*, 7 (2), 169–89.

Jha, A. (2011) Look, no embryos! The future of ethical stem cells. *The Guardian*, Accessed 13 March 2012. Available from http://www. guardian.co.uk/science/2011/mar/13/ips-reprogrammed-stem-cells.

John, M. (2011) Sexing the fetus: feminist politics and method across cultures. *Positions: East Asia Cultures Critique*, 19 (1), 7–29.

Johnston, J. (2010) America's stem cell mess. *The Scientist*, 24 (10), 33–34.

Jonas-Simpson, C. and McMahon, E. (2005) The language of loss when a baby dies prior to birth: on creating human experience. *Nursing Science Quarterly*, 18 (2), 124–30.

Karatas, J., Barlow-Stewart, K., Strong, K., Meiser, B., McMahon, C. and Roberts, C. (2010) Women's experience of pre-implantation genetic diagnosis: a qualitative study. *Prenatal Diagnosis*, 30 (8), 771–77.

Karpin, I. (1992) Legislating the female body: reproductive technology and the reconstructed woman. *Columbia Journal of Gender and Law*, 3 (1), 325–48.

DOI: 10.1057/9781137310729

Karpin, I. (2006) The uncanny embryos: legal limits to the human and reproduction without women. *Sydney Law Review*, 28, 599–623.

Karpin, I. (2010) Taking care of the 'health' of preconceived human embryos or constructing legal harms. In J. Nisker, F. Baylis, I. Karpin, C. McLeod and R. Mykitiuk (eds) *The 'Healthy' Embryo: Social, Biomedical, Legal and Philosophical Perspectives.* Cambridge: Cambridge University Press, 136–49.

Kato, M. and Sleeboom-Faulkner, M. (2011) Meanings of the embryo in Japan: narratives of IVF experience and embryo ownership. *Sociology of Health & Illness*, 33 (3), 434–47.

Kaufman, S. and Morgan, L. (2005) The anthropology of the beginnings and ends of life. *Annual Review of Anthropology*, 34, 317–41.

Kay, J. (2012) Exceptions don't work: what the Irish abortion tragedy means for the United States. Accessed 16 January 2013. Available from http://www.slate.com/blogs/xx_factor/2012/11/28/savita_halappanavar_abortion_case_what_the_irish_tragedy_means_for_the_u.html.

Keane, H. (2009) Foetal personhood and representations of the absent child in pregnancy loss memorialization. *Feminist Theory*, 10 (2), 153–71.

Keenan, J. and Stapleton, H. (2010) Bonny babies? Motherhood and nurturing the age of obesity. *Health, Risk & Society*, 12 (4), 369–83.

Kelly, S. (2012) The maternal-foetal interface and gestational chimerism: the emerging importance of chimeric bodies. *Science as Culture*, 21 (2), 233–57.

Kent, J. (2008) The fetal tissue economy: from the abortion clinic to the stem cell laboratory. *Social Science & Medicine*, 67 (11), 1747–56.

Kevin, C. (2011) 'I did not lose my baby ... my baby just died': twenty-first-century discourses of miscarriage in political and historical context. *South Atlantic Quarterly*, 110 (4), 849–65.

Kirejczyk, M. (2008) On women, egg cells and embryos. *European Journal of Women's Studies*, 15 (4), 377–91.

Kirkman, M., Rowe, H., Hardiman, A. and Rosenthal, D. (2011) Abortion is a difficult solution to a problem: a discursive analysis of interviews with women considering or undergoing abortion in Australia. *Women's Studies International Forum*, 34 (2), 121–29.

Kitzinger, J. and Williams, C. (2005) Forecasting science futures: legitimising hope and calming fears in the embryo stem cell debate. *Social Science & Medicine*, 61 (3), 731–40.

DOI: 10.1057/9781137310729

Kristeva, J. (1982) *Powers of Horror: An Essay on Abjection*. New York: Columbia University Press.

Kroløkke, C. (2010) On a trip to the womb: biotourist metaphors in fetal ultrasound imaging. *Women's Studies in Communication*, 33 (2), 138–53.

Kroløkke, C. (2011) Biotourist performances: doing parenting during the ultrasound. *Text and Performance Quarterly*, 31 (1), 15–36.

Krones, T., Schlüter, E., Neuwohner, E., El Ansari, S., Wissner, T. and Richter, G. (2006) What is the preimplantation embryo? *Social Science & Medicine*, 63 (1), 1–20.

Kukla, R. (2005) Pregnant bodies as public spaces. In S. Hardy and C. Wiedmer (eds) *Motherhood and Space: Configurations of the Maternal Through Politics, Home, and the Body*. New York: Palgrave Macmillan, 283–305.

Kukla, R. (2010) The ethics and cultural politics of reproductive risk warnings: a case study of California's Proposition 65. *Health, Risk & Society*, 12 (4), 323–34.

Kumar, A., Hessini, L. and Mitchell, E.M.H. (2009) Conceptualising abortion stigma. *Culture, Health & Sexuality*, 11 (6), 625–39.

Lalor, J., Begley, C. and Galavan, E. (2009) Recasting hope: a process of adaptation following fetal anomaly diagnosis. *Social Science & Medicine*, 68 (3), 462–72.

Landecker, H. (2011) Food as exposure: nutritional epigenetics and the new metabolism. *BioSocieties*, 6 (2), 167–94.

Langstrup, H. and Sommerlund, J. (2008) Who has more life? Authentic bodies and the ethopolitics of stem cells. *Configurations*, 16 (3), 379–98.

Lanik, A.D. (2012) Preconception counseling. *Primary Care*, 39 (1), 1–16.

Law, J.M. and Sasson, V. (2009) Introduction: restoring nuance to imaging the fetus. In J.M. Law and V. Sasson (eds) *Imaging the Fetus: The Unborn in Myth, Religion, and Culture*. New York: Oxford University Press, 3–10.

Layne, L. (2000) 'He was a real baby with baby things'. *Journal of Material Culture*, 5 (3), 321–45.

Layne, L. (2003a) *Motherhood Lost: A Feminist Account of Pregnancy Loss in America*. New York, NY: Routledge.

Layne, L. (2003b) Unhappy endings: a feminist reappraisal of the women's health movement from the vantage of pregnancy loss. *Social Science & Medicine*, 56 (9), 1881–91.

DOI: 10.1057/9781137310729

Lee, E. (2000) Young people's attitudes to abortion for abnormality. *Feminism & Psychology*, 10 (3), 396–99.

Lee, N. and Motzkau, J. (2011) Navigating the bio-politics of childhood. *Childhood*, 18 (1), 7–19.

Lennart Nilsson Photography (2013) Accessed 5 January 2013. Available from http://www.lennartnilsson.com/home.html.

Letherby, G. (1993) The meanings of miscarriage. *Women's Studies International Forum*, 16 (2), 165–80.

Littlewood, J. (1999) From the invisibility of miscarriage to an attribution of life. *Anthropology & Medicine*, 6 (2), 217–30.

Longhurst, R. (1997) 'Going nuts': re-presenting pregnant women. *New Zealand Geographer*, 53 (2), 34–39.

Longhurst, R. (2000a) *Bodies: Exploring Fluid Boundaries*. London: Routledge.

Longhurst, R. (2000b) 'Corporeographies' of pregnancy: 'bikini babes'. *Environment and Planning D: Society and Space*, 18 (4), 453–72.

Longhurst, R. (2005) *Maternities: Gender, Bodies and Space*. London: Routledge.

Lorber, J. (1987) In vitro fertilization and gender politics. *Women and Health*, 13 (1–2), 117–33.

Lowe, P. and Lee, E. (2010) Advocating alcohol abstinence to pregnant women: some observations about British policy. *Health, Risk & Society*, 12 (4), 301–11.

Lundquist, C. (2008) Being torn: toward a phenomenology of unwanted pregnancy. *Hypatia*, 23 (3), 136–55.

Lupton, D. (1995) *The Imperative of Health: Public Health and the Regulated Body*. London: Sage.

Lupton, D. (1999) Risk and the ontology of pregnant embodiment. In D. Lupton (ed) *Risk and Sociocultural Theory: New Directions and Perspectives*. Cambridge: Cambridge University Press, 59–85.

Lupton, D. (2008) 'You feel so responsible': Australian mothers' concepts and experiences related to promoting the health and development of their young children. In H. Zoller and M. Dutta (eds) *Emerging Perspectives in Health Communication*. New York: Routledge, 113–28.

Lupton, D. (2011) 'The best thing for the baby': Mothers' concepts and experiences related to promoting their infants' health and development. *Health, Risk & Society*, 13 (7–8), 637–51.

DOI: 10.1057/9781137310729

Lupton, D. (2012a) 'Precious cargo': risk and reproductive citizenship. *Critical Public Health*, 22 (3), 329–40.

Lupton, D. (2012b) The 'royal foetus' as configured via the Internet. Accessed 7 December 2012. Available from http://storify.com /DALupton/cultural-portrayals-of-the-royal-foetus.

Lupton, D. (2013a) Infant embodiment and interembodiment: a review of sociocultural perspectives. *Childhood*, 20 (1), 37–50.

Lupton, D. (2013b) Precious, pure, uncivilised, vulnerable: infants in the Australian popular media. *Children & Society*, early view online.

Lupton, D. and Schmied, V. (2013) Splitting bodies/selves: women's concepts of embodiment at the moment of birth. *Sociology of Health & Illness*, early view online.

Lyerly, A.D., Mitchell, L., Armstrong, E., Harris, L., Kukla, R., Kupperman, M. and Little, M. (2009) Risk and the pregnant body. *Hastings Center Report*, 39 (6), 34–42.

Maher, J. (2002) Visibly pregnant: toward a placental body. *Feminist Review*, (72), 95–107.

Major, B., Appelbaum, M., Beckman, L., Dutton, M., Russo, N. and West, C. (2009) Abortion and mental health: evaluating the evidence. *The American Psychologist*, 64 (9), 863–90.

Major, B., Cozzarelli, C., Cooper, M., Zubek, J., Richards, C., Wilhite, M. and Gramzow, R. (2000) Psychological responses of women after first-trimester abortion. *Archives of General Psychiatry*, 57 (8), 777–84.

Marcus, G. (2006) Assemblage. *Theory, Culture & Society*, 23 (2–3), 101–6.

Markens, S., Browner, C. and Press, N. (1997) Feeding the fetus: on interrogating the notion of maternal–fetal conflict. *Feminist Studies*, 23 (2), 351–72.

Marsland, R. and Prince, R. (2012) What is life worth? Exploring biomedical interventions, survival, and the politics of life. *Medical Anthropology Quarterly*, 26 (4), 453–69.

Martin, A. (2010) Microchimerism in the mother(land): blurring the borders of body and nation. *Body & Society*, 16 (3), 23–50.

Martin, E. (1992) *The Woman in the Body: A Cultural Analysis of Reproduction*. Boston, MA: Beacon Press.

Mason, C. (2000) Cracked babies and the partial birth of a nation: millennialism and fetal citizenship. *Cultural Studies*, 14 (1), 35–60.

McCabe, J. and Holmes, D. (2011) Reversing Kristeva's first instance of abjection: the formation of self reconsidered. *Nursing Inquiry*, 18 (1), 77–83.

DOI: 10.1057/9781137310729

McCoyd, J.L.M. (2007) Pregnancy interrupted: loss of a desired pregnancy after diagnosis of fetal anomaly. *Journal of Psychosomatic Obstetrics & Gynecology*, 28 (1), 37–48.

McCoyd, J.L.M. (2009) What do women want? Experiences and reflections of women after prenatal diagnosis and termination for anomaly. *Health Care for Women International*, 30 (6), 507–35.

McCreight, B. (2004) A grief ignored: narratives of pregnancy loss from a male perspective. *Sociology of Health & Illness*, 26 (3), 326–50.

McGinn, W. (2011) *The Meaning of Disgust*. New York: Oxford University Press.

McLeod, C. and Baylis, F. (2006) Feminists on the inalienability of human embryos. *Hypatia*, 21 (1), 1–14.

McMahon, C., Tennant, C., Ungerer, J. and Saunders, D. (1999) 'Don't count your chickens': A comparative study of the experience of pregnancy after IVF conception. *Journal of Reproductive and Infant Psychology*, 17 (4), 345–56.

McNaughton, D. (2011) From the womb to the tomb: obesity and maternal responsibility. *Critical Public Health*, 21 (2), 179–90.

Metzler, I. (2007) 'Nationalizing embryos': the politics of human embryonic stem cell research in Italy. *BioSocieties*, 2 (4), 413–27.

Michelle, C. (2006) Transgressive technologies? Strategies of discursive containment in the representation and regulation of assisted reproductive technologies in Aotearoa/New Zealand. *Women's Studies International Forum*, 29 (2), 109–24.

Mills, C. (2008) Images and emotion in abortion debates. *The American Journal of Bioethics*, 8 (12), 61–62.

Misbehaving Mums to Be (2012) Accessed 28 May 2012. Available from http://www.bbc.co.uk/programmes/b010h79q.

Mitchell, L. (2001) *Baby's First Picture: Ultrasound and the Politics of Fetal Subjects*. Toronto: University of Toronto Press.

Mitchell, L. (2004) Women's experiences of unexpected ultrasound findings. *Journal of Midwifery & Women's Health*, 49 (3), 228–34.

Mitchell, L. and Georges, E. (1997) Cross-cultural cyborgs: Greek and Canadian women's discourses on fetal ultrasound. *Feminist Studies*, 23 (2), 373–401.

Mitchell, R. and Waldby, C. (2006) *Tissue Economies: Blood, Organs, and Cell Lines in Late Capitalism*. Durham: Duke University Press.

Mobile Ultrasounds (2012) Accessed 17 September 2012. Available from http://www.mobileultrasounds.com.au/information.html.

DOI: 10.1057/9781137310729

Mohapatra, S. (2012) Stateless babies and adoption scams: a bioethical analysis of international commercial surrogacy. *Berkeley Journal of International Law*, 30 (2), 412–50.

Morgan, L. (1996) Fetal relationality in feminist philosophy: an anthropological critique. *Hypatia*, 11 (3), 47–70.

Morgan, L. (1997) Imagining the unborn in the Ecuadorian Andes. *Feminist Studies*, 23 (2), 323–50.

Morgan, L. (2002) 'Properly disposed of': a history of embryo disposal and the changing claims on fetal remains. *Medical Anthropology*, 21 (3–4), 247–74.

Morgan, L. (2006a) 'Life begins when they steal your bicycle': cross-cultural practices of personhood at the beginnings and ends of life. *The Journal of Law, Medicine & Ethics*, 34 (1), 8–15.

Morgan, L. (2006b) The rise and demise of a collection of human fetuses at Mount Holyoke College. *Perspectives in Biology and Medicine*, 49 (3), 435–51.

Morgan, L. (2006c) Strange anatomy: Gertrude Stein and the avant-garde embryo. *Hypatia*, 21 (1), 15–34.

Morgan, L. (2009) *Icons of Life: A Cultural History of Human Embryos*. Berkeley, CA: University of California Press.

Mulkay, M. (1994) Embryos in the news. *Public Understanding of Science*, 3 (1), 33–51.

Mulkay, M. (1996) Frankenstein and the debate over embryo research. *Science, Technology, & Human Values*, 21 (2), 157–76.

The Multi Dimensional Human Embryo (2009) Accessed 10 April 2012. Available from http://embryo.soad.umich.edu/.

Nash, M. (2007) From 'bump' to 'baby': gazing at the foetus in 4D. *Philament*, Accessed 3 December 2012. Available from http://sydney.edu.au/arts/publications/philament/issue10_contents.htm.

Nash, M. (2011) 'You don't train for a marathon sitting on the couch': performances of pregnancy 'fitness' and 'good' motherhood in Melbourne, Australia. *Women's Studies International Forum*, 34 (1), 50–65.

Nash, M. (2013) *Making 'Postmodern' Mothers: Pregnant Embodiment, Baby Bumps and Body Image*. Houndmills: Palgrave Macmillan.

Neiterman, E. (2012) Doing pregnancy: pregnant embodiment as performance. *Women's Studies International Forum*, 35 (5), 372–83.

Newman, K. (1996) *Fetal Positions: Individualism, Science, Individuality,*. Stanford, CA: Stanford University Press.

DOI: 10.1057/9781137310729

Oakley, A. (1984) *The Captured Womb: A History of Medical Care of Pregnant Women*. Oxford: Basil Blackwell.

Obstetric Ultrasound: A Comprehensive Guide (2013) Accessed 20 January 2013. Available from http://www.ob-ultrasound.net/.

Palmer, J. (2009a) The placental body in 3D: everyday practices of non-diagnostic sonography. *Feminist Review*, 93, 64–80.

Palmer, J. (2009b) Seeing and knowing: ultrasound images in the contemporary abortion debate. *Feminist Theory*, 10 (2), 173–89.

Paltrow, L.M. and Flavin, J. (2013) Arrests of and forced interventions on pregnant women in the United States, 1973–2005: implications for women's legal status and public health. *Journal of Health Politics, Policy and Law*, 38 (2), 299–343.

Pande, A. (2010) Commercial surrogacy in India: manufacturing a perfect mother–worker. *Signs*, 35 (4), 969–92.

Parkes, G. (1999) European section. *Journal of Social Welfare and Family Law*, 21 (3), 285–294.

Parry, S. (2006) (Re)constructing embryos in stem cell research: exploring the meaning of embryos for people involved in fertility treatments. *Social Science & Medicine*, 62 (10), 2349–59.

Perrett, R.W. (2000) Buddhism, abortion and the middle way. *Asian Philosophy*, 10 (2), 101–14.

Petchesky, R. (1980) Reproductive freedom: beyond a 'woman's right to choose'. *Signs*, 5 (4), 661–85.

Petchesky, R. (1987) Fetal images: the power of visual culture in the politics of reproduction. *Feminist Studies*, 13 (2), 263–92.

Petersen, A. (2002) Replicating our bodies, losing our selves: news media portrayals of human cloning in the wake of Dolly. *Body & Society*, 8 (4), 71–90.

Petersen, A. (2005) The metaphors of risk: biotechnology in the news. *Health, Risk & Society*, 7 (3), 203–8.

Pfeffer, N. (2008) What British women say matters to them about donating an aborted fetus to stem cell research: a focus group study. *Social Science & Medicine*, 66 (12), 2544–54.

Pfeffer, N. (2009) How work reconfigures an 'unwanted' pregnancy into 'the right tool for the job' in stem cell research. *Sociology of Health & Illness*, 31 (1), 98–111.

Pickert, K. (2013) Has the fight for abortion rights been lost? Accessed 8 January 2013. Available from http://ideas.time.com/2013/01/03/has-the-fight-for-abortion-rights-been-lost/.

DOI: 10.1057/9781137310729

Pickett, K.E., Wakschlag, L.S., Rathouz, P.J., Leventhal, B.L. and Abrams, B. (2002) The working-class context of pregnancy smoking. *Health & Place*, 8 (3), 167–75.

Prainsack, B., Geesink, I. and Franklin, S. (2008) Stem cell technologies 1998–2008: controversies and silences. *Science as Culture*, 17 (4), 351–62.

Prangell, T. 2011. Protecting precious cargo. *The Sun-Herald* (special supplement on pregnancy), 24 July 2011, Supplement 1.

Ralph, S., Morris, M., Fallows, S. and Abayomi, J.C. (2011) Does maternal early pregnancy underweight influence neonatal birth outcome? A retrospective study in Liverpool. *Proceedings of the Nutrition Society*, (OCE1). Accessed 15 December 2012. Available from http://journals.cambridge.org/action/displayFulltext?type=6&fid=83 47012&jid=PNS&volumeId=70&issueId=OCE1&aid=8347011&fulltex tType=AB&fileId=S0029665111000267.

Rapp, R. (1990) Constructing amniocentesis: maternal and medical discourses. In F. Ginsburg and A.L. Tsing (eds) *Uncertain Terms: Negotiating Gender in American Culture*. Boston, MA: Beacon Press, 28–42.

Rapp, R. (2000) *Testing Women, Testing the Fetus: The Social Impact of Amniocentesis in America*. New York: Routledge.

Remennick, L. (2007) The quest for the perfect baby: why do Israeli women seek prenatal genetic testing? *Sociology of Health & Illness*, 28 (1), 21–53.

Rigdon, S. (1996) Abortion law and practice in China: an overview with comparisons to the United States. *Social Science & Medicine*, 42 (4), 543–60.

Roberts, C. and Franklin, S. (2004) Experiencing new forms of genetic choice: findings from an ethnographic study of preimplantation genetic diagnosis. *Human Fertility*, 7 (4), 285–93.

Roberts, J. (2012) 'Wakey wakey baby': narrating four-dimensional (4-D) bonding scans. *Sociology of Health & Illness*, 34 (2), 219–314.

Roberts, N. and Thilaganathan, B. (2007) The role of ultrasound in obstetrics. *Obstetrics, Gynaecology & Reproductive Medicine*, 17 (3), 79–85.

Robertson, J.A. (1995) The case of the switched embryos. *The Hastings Center Report*, 25 (6), 13–19.

Rochman, B. (2012) Is Kate Middleton too thin to be pregnant? *Time*, Accessed 7 December 2012. Available from http://healthland.time.com/2012/12/07/is-kate-middleton-too-thin-to-be-pregnant/.

DOI: 10.1057/9781137310729

Root, R. and Browner, C. (2001) Practices of the pregnant self: compliance with and resistance to prenatal norms. *Culture, Medicine, and Psychiatry*, 25, 196–223.

Rose, N. (2007a) Molecular biopolitics, somatic ethics and the spirit of biocapital. *Social Theory & Health*, 5 (1), 3–29.

Rose, N. (2007b) *The Politics of Life Itself. Biomedicine, Power, and Subjectivity in the Twenty-First Century*. Princeton, NJ: Princeton University Press.

Rose, N. (2009) Normality and pathology in a biomedical age. *The Sociological Review*, 57 (s2), 66–83.

Rothman, B.K. (1994) *The Tentative Pregnancy: Amniocentesis and the Sexual Politics of Motherhood*. 2nd edn, London: Pandora.

Rowland, R. (1992) *Living Laboratories: Women and Reproductive Technologies*. Sydney: Pan Macmillan.

Rubin, B. (2008) Therapeutic promise in the discourse of human embryonic stem cell research. *Science as Culture*, 17 (1), 13–27.

Ruddick, S. (2007) At the horizons of the subject: neo-liberalism, neo-conservatism and the rights of the child Part One: from 'knowing' fetus to 'confused' child. *Gender, Place & Culture*, 14 (5), 513–27.

Ruhl, L. (1999) Liberal governance and prenatal care: risk and regulation in pregnancy. *Economy and Society*, 28 (1), 95–117.

Ruhl, L. (2002) Disarticulating liberal subjectivities: abortion and fetal protection. *Feminist Studies*, 28 (1), 37–60.

Rylko-Bauer, B. (1996) Abortion from a crosscultural perspective: an introduction. *Social Science & Medicine*, 42 (4), 479–82.

Salmon, A. (2011) Aboriginal mothering, FASD prevention and the contestations of neoliberal citizenship. *Critical Public Health*, 21 (2), 165–78.

Sandelowski, M. (1994) Separate, but less unequal: fetal ultrasonography and the transformation of expectant mother/fatherhood. *Gender & Society*, 8 (2), 230–45.

Savulescu, J. (2001) Procreative beneficence: why we should select the best children. *Bioethics*, 15 (5–6), 413–26.

Savulescu, J. (2012) 'Liberals are disgusting': in defence of the publication of 'After-birth Abortion'. *Journal of Medical Ethics Blog*, Accessed 24 March 2012. Available from http://blogs.bmj.com. ezproxy1.library.usyd.edu.au/medical-ethics/2012/02/28 /liberals-are-disgusting-in-defence-of-the-publication-of-after-birth-abortion/.

DOI: 10.1057/9781137310729

Schmied, V. and Lupton, D. (2001) The externality of the inside: body images of pregnancy. *Nursing Inquiry*, 8 (1), 32–40.

Scully, J.L., Banks, S. and Shakespeare, T. (2006a) Chance, choice and control: lay debate on prenatal social sex selection. *Social Science & Medicine (1982)*, 63 (1), 21–31.

Scully, J.L., Shakespeare, T. and Banks, S. (2006b) Gift not commodity? Lay people deliberating social sex selection. *Sociology of Health & Illness*, 28 (6), 749–67.

Shakespeare, T. (1999) Manifesto for genetic justice. *Social Alternatives*, 18 (1), 29–32.

Sharp, K. and Earle, S. (2002) Feminism, abortion and disability: irreconcilable differences? *Disability & Society*, 17 (2), 137–45.

Shaw, R. and Giles, D. (2009) Motherhood on ice? A media framing analysis of older mothers in the UK news. *Psychology & Health*, 24 (2), 221–36.

Shellenberg, K.M., Moore, A.M., Bankole, A., Juarez, F., Omideyi, A.K., Palomino, N., Sathar, Z., Singh, S. and Tsui, A.O. (2011) Social stigma and disclosure about induced abortion: results from an exploratory study. *Global Public Health*, 6 (S1), S111–S125.

Sherwin, S. (1992) *No Longer Patient: Feminist Ethics and Health Care*. Philadelphia: Temple University Press.

Shetty, P. (2012) India's unregulated surrogacy industry. *The Lancet*, 380 (9854), 1633–34.

Shildrick, M. (1997) *Leaky Bodies and Boundaries: Feminism, Postmodernism and (Bio)ethics*. London: Routledge.

Shrage, L. (2002) From reproductive rights to reproductive Barbie: post-porn modernism and abortion. *Feminist Studies*, 28 (1), 61–93.

Smith, J.L. (2003) 'Suitable mothers': lesbian and single women and the 'unborn' in Australian parliamentary discourse. *Critical Social Policy*, 23 (1), 63–88.

Smyth, L. (2002) Feminism and abortion politics. *Women's Studies International Forum*, 25 (3), 335–45.

Snowflakes Embryo Adoption (2012) Accessed 9 December 2012. Available from http://www.nightlight.org/snowflake-embryo-adoption/.

The Sociology of the Unborn (2013) Accessed 6 January 2013. Available from http://pinterest.com/dalupton/sociology-of-the-unborn.

Solodnikov, V.V. (2010) Abortion. *Sociological Research*, 49 (5), 74–96.

Sperling, S. (2008) Converting ethics into reason: German stem cell policy between science and the law. *Science as Culture*, 17 (4), 363–75.

DOI: 10.1057/9781137310729

Stabile, C.A. (1992) Shooting the mother: fetal photography and the politics of disappearance. *Camera Obscura*, 10 (1), 178–205.

Star, S.L. and Grisemer, J. (1989) Institutional ecology, 'translations' and boundary objects: amateurs and professionals in Berkeley's Museum of Vertebrate Zoology. *Social Studies of Science*, 19 (3), 387–420.

Steinbock, B. (2000) What does 'respect for embryos' mean in the context of stem cell research? *Women's Health Issues*, 10 (3), 127–30.

Stephens, J. (2005) Cultural Memory, feminism and motherhood. *Arena Journal*, (24), 69–82.

STFU Parents (2011) Parents on fetuses using social media. Accessed 28 September 2012. Available from http://www.mommyish.com/2011/06/21/stfu-parents-parents-on-fetuses-using-social-media/.

STFU Parents (2012) Are ultrasound photos still even considered Facebook overshare? Accessed 16 December 2012. Available from http://www.mommyish.com/2012/12/14/stfu-parents-are-ultrasound-photos-still-even-considered-overshare/.

Stormer, N. (2008) Looking in wonder: prenatal sublimity and the commonplace 'life'. *Signs*, 33 (3), 647–73.

Svendsen, M. (2007) Between reproductive and regenerative medicine: practising embryo donation and civil responsibility in Denmark. *Body & Society*, 13 (4), 21–45.

Svendsen, M. and Koch, L. (2008) Unpacking the 'spare embryo'. *Social Studies of Science*, 38 (1), 93–110.

Taylor, J. (1992) The public fetus and the family car: from abortion politics to a Volvo advertisement. *Public Culture*, 4 (2), 67–80.

Taylor, J. (2000) Of sonograms and baby prams: prenatal diagnosis, pregnancy, and consumption. *Feminist Studies*, 26 (2), 391–418.

Taylor, J. (2008) *The Public Life of the Fetal Sonogram: Technology, Consumption and the Politics of Reproduction*. New Brunswick, NJ: Rutgers University Press.

Teman, E. (2001) Technological fragmentation and women's empowerment: surrogate motherhood in Israel. *Women's Studies Quarterly*, 29 (3/4), 11–34.

Teman, E. (2009) Embodying surrogate motherhood: pregnancy as a dyadic body-project. *Body & Society*, 15 (3), 47–69.

TheVisualMD.com (2013) Accessed 19 January 2013. Available from http://www.thevisualmd.com/.

DOI: 10.1057/9781137310729

Tsiaras, A. (2011) 'Conception to Birth – Visualized' (video). Accessed 11 August 2012. Available from http://www.youtube.com /watch?v=fKyljukBE7o.

Tsiaras, A. (2012) 'From a Cell to a Baby' (video). Accessed 22 January 2013. Available from http://www.thevisualmd.com/videos/result /from_a_cell_to_a_baby.

Tsiaras, A. (2013) The beautiful and efficient anatomy of pregnancy. Accessed 19 January 2013. Available from http://www.huffingtonpost. com/alexander-tsiaras/pregnancy-anatomy_b_2499945.html?ncid=e dlinkusaolp00000003&ir=Parents.

Tyler, I. (2009) Introduction: birth. *Feminist Review*, 93, 1–7.

The Ultrasound as Cultural Artefact (2013) Accessed 6 January 2013. Available from http://pinterest.com/dalupton/the-ultrasound-as-cultural-artefact.

Use of Human Embryos in Research (2009) Accessed 15 August 2012. Available from http://www.asscr.org/index.php?id=1031.

van Balen, F. (1998) Development of IVF children. *Developmental Review*, 18 (1), 30–46.

Van den Bergh, B. and Simons, A. (2009) A review of scales to measure the mother-foetus relationship. *Journal of Reproductive and Infant Psychology*, 27 (2), 114–26.

Van der Ploeg, I. (2001) *Prosthetic Bodies: The Construction of the Fetus and the Couple as Patients in Reproductive Technologies*. London: Kluwer.

Van der Ploeg, I. (2004) 'Only angels can do without skin': on reproductive technology's hybrids and the politics of body boundaries. *Body & Society*, 10 (2–3), 153–81.

The Visible Embryo (2012) Accessed 19 September 2012. Available from http://www.visembryo.com/baby/index.html.

The Visible Embryo Pregnancy Timeline (2011) Accessed 19 September 2012. Available from http://www.visembryo.com/baby /pregnancytimeline.html.

Vlassoff, M., Walker, D., Shearer, J., Newlands, D. and Singh, S. (2009) Estimates of health care system costs of unsafe abortion in Africa and Latin America. *International Perspectives on Sexual and Reproductive Health*, 35 (3), 114–21.

Vogel, L. (2012) Sex selection migrates to Canada. *Canadian Medical Association Journal*, 184 (3), E163–64.

DOI: 10.1057/9781137310729

Wainwright, S., Williams, C., Michael, M., Farsides, B. and Cribb, A. (2006) Ethical boundary-work in the embryonic stem cell laboratory. *Sociology of Health & Illness*, 28 (6), 732–38.

Wainwright, S.P., Michael, M. and Williams, C. (2008) Shifting paradigms? Reflections on regenerative medicine, embryonic stem cells and pharmaceuticals. *Sociology of Health & Illness*, 30 (6), 959–74.

Waldby, C. (2002a) Biomedicine, tissue transfer and intercorporeality. *Feminist Theory*, 3 (3), 239–54.

Waldby, C. (2002b) Stem cells, tissue cultures and the production of biovalue. *Health*, 6 (3), 305–23.

Waldby, C. (2008) Oocyte markets: women's reproductive work in embryonic stem cell research. *New Genetics and Society*, 27 (1), 19–31.

Waldby, C. and Carroll, K. (2012) Egg donation for stem cell research: ideas of surplus and deficit in Australian IVF patients' and reproductive donors' accounts. *Sociology of Health & Illness*, 34 (4), 513–28.

Waldby, C. and Cooper, M. (2010) From reproductive work to regenerative labour: the female body and the stem cell industries. *Feminist Theory*, 11 (3), 3–22.

Waldby, C. and Squier, S.M. (2003) Ontogeny, ontology, and phylogeny: embryonic life and stem cell technologies. *Configurations*, 11 (1), 27–46.

Warin, M., Zivkovic, T., Moore, V. and Davies, M. (2012) Mothers as smoking guns: fetal overnutrition and the reproduction of obesity. *Feminism & Psychology*, 22 (3), 360–75.

Warren, M.A. (1988) IVF and women's interests: an analysis of feminist concerns. *Bioethics*, 2 (1), 37–57.

Warren, M.A. (1989) The moral significance of birth. *Hypatia*, 4 (3), 46–65.

Warren, S. and Brewis, J. (2004) Matter over mind? Examining the experience of pregnancy. *Sociology*, 38 (2), 219–36.

Weir, L. (1996) Recent developments in the government of pregnancy. *Economy and Society*, 25 (3), 373–92.

Weir, L. (1998) Cultural intertexts and scientific rationality: the case of pregnancy ultrasound. *Economy and Society*, 27 (2–3), 249–53.

Weir, L. (2006) *Pregnancy, Risk, and Biopolitics: On the Threshold of the Living Subject*. London: Routledge.

Williams, C. (2005) Framing the fetus in medical work: rituals and practices. *Social Science & Medicine*, 60 (9), 2085–95.

DOI: 10.1057/9781137310729

Williams, C. (2006) Dilemmas in fetal medicine: premature application of technology or responding to women's choice? *Sociology of Health & Illness*, 28 (1), 1–20.

Williams, C., Kitzinger, J. and Henderson, L. (2003) Envisaging the embryo in stem cell research: rhetorical strategies and media reporting of the ethical debates. *Sociology of Health & Illness*, 25 (7), 793–814.

Williams, C., Sandall, J., Lewando-Hundt, G., Heyman, B., Spencer, K. and Grellier, R. (2005) Women as moral pioneers? Experiences of first trimester antenatal screening. *Social Science & Medicine*, 61 (9), 1983–92.

Williams, C., Wainwright, S., Ehrich, K. and Michael, M. (2008) Human embryos as boundary objects? Some reflections on the biomedical worlds of embryonic stem cells and pre-implantation genetic diagnosis. *New Genetics and Society*, 27 (1), 7–18.

Woo, J. (2002) A short history of the development of ultrasound in obstetrics and gynecology. Accessed 16 November 2012. Available from http://www.ob-ultrasound.net/history1.html.

The World's Abortion Laws (2013) Accessed 21 January 2013. Available from http://worldabortionlaws.com/questions.html.

Young, I.M. (1984) Pregnant embodiment: subjectivity and alienation. *Journal of Medicine and Philosophy*, 9 (1), 45–62.

Zelizer, V. (1985) *Pricing the Priceless Child: The Changing Social Value of Children*. Princeton, NJ: Princeton University Press.

Zheng, X., Pang, L., Tellier, S., Tan, L., Zhang, L., Hu, Y. and Wei, J. (2013) The changing patterns of abortion among married women in China, 1984–2005. *European Journal of Obstetrics & Gynecology and Reproductive Biology*, 166 (1), 70–75.

DOI: 10.1057/9781137310729

Index

DOI: 10.1057/9781137310729

DOI: 10.1057/9781137310729

DOI: 10.1057/9781137310729

DOI: 10.1057/9781137310729

DOI: 10.1057/9781137310729

CPSIA information can be obtained at www.ICGtesting.com
Printed in the USA
LVOW13*1503280314

379388LV00007B/118/P